SEMISYNTHETIC PROTEINS

SEMISYNTHETIC PROTEINS

R. E. Offord
University Lecturer in Molecular Biophysics and Tutor in Biochemistry, Christ Church, Oxford

A Wiley-Interscience Publication

JOHN WILEY & SONS LTD
Chichester · New York · Brisbane · Toronto

British Library Cataloguing in Publication Data:

Offord, R E

Semisynthetic proteins.
1. Proteins
2. Synthetic products

I. Title

547′.75 QD431 79-40521

ISBN 0 471 27615 4

PRINTED IN GREAT BRITAIN BY JOHN WRIGHT & SONS LTD.,
AT THE STONEBRIDGE PRESS, BRISTOL BS4 5NU

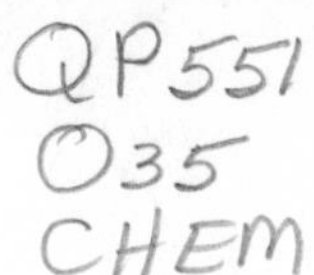

Contents

Preface

I have written this book, which I hope will be read by biochemists, molecular biologists, and pharmacologists as well as by organic chemists, because I believe that the semisynthetic approach is at present the most generally useful means of obtaining proteins that have structures not to be found in nature. Some indication of the range of work now being carried out is given by the selection of Figures that conclude Appendix 1 of the book: a detailed review of current developments is given in Chapter 8. Already, if we restrict ourselves to molecules larger than insulin, there have been more protein analogues produced by semisynthesis than by traditional methods.

I believe that the time has come to collect the previously scattered descriptions of the techniques of protein semisynthesis into one place, since these techniques are now at a state in which they can be used on a wide range of problems with an excellent chance of a successful outcome. I have not sought to conceal the remaining difficulties and imperfections and I hope that what I have written might lead to their more rapid elimination. Most of all, however, I hope to reach those who have a need for a protein analogue but who perhaps never thought themselves likely to undertake the chemical synthesis of a protein. I hope that having read this book, some at least will make the attempt.

I have included a great many practical Examples in the text. Many of them are drawn from procedures that originated in this Laboratory, or were further developed here. Some of these are previously unpublished, and those that have been published before now include, where appropriate, our latest alterations to points of detail. I have also been able to use the Notes to the Examples to make a more extended commentary on practical details than a Journal can usually allow. Other Examples are based on publications of other laboratories, although nearly all have been used here for one purpose or another. As given in this book, such methods often include minor modifications that we have made, usually to suit our own, local convenience. I have made it clear when this has been done. These modifications should not be taken to imply that the original method was unsatisfactory and the reader might well prefer to follow the original procedures. I simply felt that I could not comment on them in a similar style to that used on our own methods unless they were in a form with which I was personally familiar.

The practical details have usually been checked by two or more people with experience of the method described, with the typescript having been read by at least one of them against working notes of the preparation. Should, in spite of this, errors have crept in, I would be most grateful to be informed.

I would like to say how much I have appreciated the friendly reception of my ideas given by those working on the total, organic-chemical synthesis of peptides and proteins: in particular by the participants and organizers of the European Peptide Symposia. In recent years they have suffered the untimely loss from their number of Ernesto Scoffone, Josef Rudinger and George Kenner—I had counted on the guidance of all three in writing this book.

R. E. OFFORD

Acknowledgments

I am deeply grateful to the Medical Research Council, and to Professor D. C. Phillips, for their support and encouragement throughout the studies that led to the preparation of this book.

I owe a very great deal to those colleagues who have worked with me, either part- or full-time, on semisynthetic projects in the last few years: F. Borrás-Cuesta; C. G. Bradshaw; J. G. Davies; P. A. Halban; D. E. Harris, C. Herrera-Caravajal; W. H. Johnson; P. C. Mahony; J. D. Milman; R. R. Moore; A. R. Rees; D. J. Saunders; H. T. Storey; C. J. A. Wallace; D. Webster; and J. Welford. Also, many of them have kindly read over those sections of the text that concern them and in particular checked the appropriate Examples with scrupulous care. D. E. Harris in addition read the text as a whole and closely guided my writing of the sections on non-covalent associations. I did not always take the advice that they offered me and remain responsible for all errors and infelicities. I would also like to thank the undergraduates who have carried out projects here as a part of their first-degree courses: T. A. Backer, N. Bate; S. J. Cox; J. M. Dixon; C. F. Hayward; S. Hoare; R. S. Moore; D. E. Neuberger; M. H. Pheasey; and D. A. Wightman.

Many colleagues in other laboratories in the U.K. and abroad have helped me very much. I have particularly appreciated the advice of Dr. G. T. Young.

I thank the authors and publishers who have given permission for the use of copyright material for Figures and Tables, and who are named in the Legends.

Mrs Susan Friend typed both the draft and final versions of the text: I am indebted to her for her skill and enthusiasm. I must also thank Mrs. V. M. Crean for her extremely helpful copy-editing.

ACKNOWLEDGMENTS

Abbreviations

Acim- = acetimidyl-

Boc- = t-butyloxycarbonyl-

Cbz- = benzyloxycarbonyl-

Msc- = methylsulphonylethyloxycarbonyl-

The remaining abbreviations are those permitted to be used without definition by the Commission on Biochemical Nomenclature (see *Biochem. J.* (1966) **101,** 1–7).

CHAPTER 1

The synthesis and semisynthesis of proteins

This book describes the technique of protein semisynthesis, in which fragments of naturally occurring proteins are used as ready-made intermediates in the construction of proteins of novel structure. Modern biology and medicine have produced a great demand for such proteins, and the development of simple and generally applicable methods of obtaining them is a worthwhile aim. In the laboratory, there is a need for molecules of altered sequence, in order to be able to observe the effect of the alterations on biological activity. The alteration could be the substitution of a particularly important residue by another of the 20 common types or by a novel one designed for the purpose. Still in the laboratory, one might wish to introduce groups or isotopic substitutions that would confer novel spectroscopic properties (visible-light, ultraviolet, fluorescence or microwave) on the molecule. Products of this type of alteration can be used to study the structure of proteins, and the dynamics of their interaction with other molecules, in ways that would otherwise be impossible. Radioactively labelled proteins are often required for tracer studies in both laboratory and clinical practice. Looking even more toward practical applications, it is becoming possible to design modifications to polypeptides that would enhance an existing property, or confer a new one. For this possibility to make any technological or medical impact either the modified molecules would have to be potent in extremely small quantities or the methods of preparing them would have to be quite simple. A similar constraint would apply to any scheme for supplying the need for a natural, but scarce, protein by modifying a more common one.

So far, these needs have been met, if at all, by side-chain modification of the intact protein by reagents that are specific for certain types of chemical group. Much of what we know about the relationship between protein structure and function rests on experiments with materials produced in this way, and this approach will continue to be very important for as far ahead as can be foreseen. However, only some of the transformations that we might require can be carried out by these means. Even in those cases in which a reagent exists that is theoretically capable of bringing the change about, it is often very difficult to restrict its action to one type of side-chain. It is still less easy, when that particular side-chain occurs more than once in the structure, to direct the reagent exclusively to the one position that we wish to alter. Reaction at several sites in the protein can sometimes be tolerated, but not always.

Chemical synthesis would appear to offer a much wider scope. Given a desired amino-acid sequence, can we not build it up, one residue at a time, from its component amino acids? Such methods have been developed during the present century and the rest of this Chapter is mainly devoted to a brief description of them. We shall see that the synthesis of such large molecules, composed as they are of units that are susceptible to racemization and many types of side-reaction, is a formidable task. Most chemical syntheses are of compounds that possess a few dozen atoms at most, and yet we now contemplate assembling molecules that have thousands.

Few but the most specialized laboratories can embark with any confidence on the synthesis of even quite small peptides, and even these can attempt only with immense effort the synthesis of the very smallest of proteins. Such syntheses that there have been represent major scientific achievements, but those that carried them out would be the first to point out that the magnitude of the task, even using the latest automated methods, has imposed severe limitations on both the quantity and the quality of the material produced. We shall see that a few syntheses have been attempted on proteins of a 100 residues or more, all molecules well below the median size of the globular proteins. The usual result, after years of the most skilled efforts of several people, is a rather small quantity of material which, though with some biological activity, is not always above suspicion in its chemical integrity. At the time of writing no variant of a polypeptide chain above 100 residues, synthesized in the manner described, has been used for any of the purposes mentioned at the beginning of this Chapter.

Even when we turn to much smaller proteins, such as insulin where there have now been several syntheses of quite satisfactory materials, the effort is huge and the quantities produced are fairly small. The 40 mg of human insulin synthesized by Sieber *et al.* (1974) may well have equalled in mass the sum of all other samples of synthetic insulin produced throughout the world up to that time. The 629 mg of their more recent preparation (Sieber *et al.*, 1977) is even more out of proportion to the normal scale of such exercises.

If we really can use the fragments of natural proteins as ready-made intermediates, then the overall process becomes much more simple. This is particularly true for those modifications in which, as is so often the case, the desired difference from the natural protein is only in a single residue, or perhaps in a single atom. The majority of the structure could then be made up from pieces of natural origin, rather than synthesizing at great effort 99% or more of the molecule as nature intended, simply to be able to introduce one minor change. One may reasonably hope that the likelihood of racemization and structural error is less in the biologically derived parts of the molecule than would be the case if they were synthetic. Fewer people are needed for the work, and they do not require quite so specialized a background. We shall see later that there have already been more semisyntheses of proteins of over 100 residues than there have been conventional* syntheses.

This is not to deny that semisynthesis has its disadvantages and difficulties. What these are, and the way that one attempts to deal with them, will become apparent in Chapters 2–7, which describe the methodology of semisynthesis. I hope that my drawing attention to the remaining disadvantages might help stimulate the work needed to overcome them. Nonetheless, it is reasonable to claim that protein semisynthesis has now been shown in practice to have most of the advantages suggested for it. It is to be hoped that the outstanding problems of conventional synthesis will be overcome, but until they are, semisynthesis offers the best hope of synthesizing usable samples of variant proteins of low molecular weight, and the only hope when the required molecule is of any size.

It is not yet clear to what extent genetic manipulation will join conventional synthesis and semisynthesis as a means of obtaining analogues of proteins. It seems very likely that it will be used to give proteins of natural sequence, and analogues that differ from them through substitutions involving the ordinary, genetically coded amino acids—although the economics of producing material in bulk could remain very unfavourable for some time to come. It is to be hoped that these developments will become practical as soon as possible, if for no other reason than to free those with chemical expertise for the task of preparing analogues that (like many of the more interesting ones) are likely to be difficult to produce biologically. Of course, even substitutions that involve the insertion in specific places of unusual, or isotopically substituted, residues could be achieved by the manipulation of the structure of coding triplets and of transfer RNA; it cannot be said when, if ever, such an approach would become sufficiently simple to be competitive with chemical methods.

It is likely that the production of semisynthetic analogues of practical value would be stimulated by the availability of really large supplies of proteins of native, or near native, sequence. In particular it is not at present easy to envisage the introduction into general use of semisynthetic insulin analogues, however useful their properties, because of the scarcity of the native hormone needed as the raw material for their production.

Before we consider how semisyntheses are designed and executed, it is necessary to survey the methodology of conventional synthesis.

THE ESSENTIAL OPERATIONS IN CONVENTIONAL PEPTIDE SYNTHESIS

Conventional peptide synthesis is an intricate art, and a critical review of its principles and present-day practice would demand a book in itself. The introduction to the Bibliography lists a number of such works. The following section

*I have, in the interests of brevity, used 'conventional' throughout this book to indicate 'other than semisynthetic'. I do not intend any of the overtones that the term carries in literary or conversational use. '*De novo*' and 'total' did not seem, strictly speaking, to be applicable—the L-amino acids used in such work are often of biological origin.

(pp. 4–22) is written primarily for those without practical experience of peptide synthesis. It presents only in outline the major issues that have to be taken into account when one designs a synthetic route. The emphasis is on those aspects that most clearly show the relationships between the conventional and semi-synthetic approaches and which introduce the terminology used in the subsequent Chapters.

Coupling

Amino and carboxyl components

Coupling, that is the condensation between the α-COOH group of one amino-acid derivative and the α-NH_2 group of another, is the distinctive operation of peptide synthesis. The molecule donating the —COOH group to the peptide bond so formed is called the carboxyl component and the molecule donating the —NH_2 is called the amino component. (Therefore, somewhat confusingly, the carboxyl component finishes up on the amino-terminal side of the peptide bond and the amino component finishes up on the carboxyl-terminal side.)

We can write the process as

$$\underset{\text{carboxyl component}}{\text{X}-\text{NH}-\overset{\overset{\displaystyle R_2}{|}}{\text{CH}}-\text{COOH}} + \underset{\text{amino component}}{\text{H}_2\text{N}-\overset{\overset{\displaystyle R_1}{|}}{\text{CH}}-\text{CO}-\text{Y}} \rightleftharpoons$$

$$\text{X}-\text{NH}-\overset{\overset{\displaystyle R_2}{|}}{\text{CH}}-\text{CO}-\text{NH}-\overset{\overset{\displaystyle R_1}{|}}{\text{CH}}-\text{CO}-\text{Y} + \text{H}_2\text{O} \qquad (1.1)$$

where X— and Y— represent the groups forming the amino-acid derivatives. We shall see in a moment that the actual chemical mechanisms employed are more complicated, principally because of the need to force the reaction to completion without using such severe conditions that side-reactions occur. It is also necessary to prevent racemization of the chiral centres of the amino acids. It is such considerations, for example, that deter the modern peptide chemist from forming the acyl chloride of the carboxyl component and reacting it directly with the amino component.

Methods of coupling: azide, carbodiimide and active ester

The azide method (Curtius, 1902, 1904) is among the oldest. The carboxyl component is esterified and then activated for coupling in the following stages

$$\text{X}-\text{NH}-\overset{\overset{\displaystyle R_2}{|}}{\text{CH}}-\text{CO}-\text{OR} + \text{H}_2\text{N}-\text{NH}_2 \rightleftharpoons$$

$$\text{X}-\text{NH}-\overset{\overset{\displaystyle R_2}{|}}{\text{CH}}-\text{CO}-\text{NH}-\text{NH}_2 + \text{ROH} \qquad (1.2)$$

$$X{-}NH{-}\underset{\displaystyle |}{\overset{R_2}{}}\!\!CH{-}CO{-}NH{-}NH_2 + HNO_2 \rightleftharpoons$$

$$X{-}NH{-}CH(R_2){-}CO{-}N_3 + 2\,H_2O \qquad (1.3)$$

The carboxyl component can be stored as the hydrazo derivative (Reaction (1.2)) without racemization or decomposition, but the azide (Reaction (1.3)) is unstable and should preferably be formed at a low temperature and then used at once, still at a low temperature.

$$X{-}NH{-}CH(R_2){-}CO{-}N_3 + H_2N{-}CH(R_1){-}CO{-}Y \rightleftharpoons$$

$$X{-}NH{-}CH(R_2){-}CO{-}NH{-}CH(R_1){-}CO{-}Y + HN_3 \qquad (1.4)$$

The azide method's principal advantage is normally thought to be that the mechanism precludes any racemization whatever. However, cases have been found, when too strongly basic conditions have been employed, when racemization is significant (Sieber and Riniker, 1973). The azides, once formed, are prone to undergo the Curtius rearrangement to an isocyanate, which can then attack certain side-chains.

$$X{-}NH{-}CH(R_2){-}CO{-}N_3 \rightleftharpoons X{-}NH{-}CH(R_2){-}N{=}C{=}O + N_2 \qquad (1.5)$$

$$X{-}NH{-}CH(CH_2OH){-}N{=}C{=}O \rightleftharpoons X{-}NH{-}\underset{\text{ring: }H_2C{-}O{-}C(=O){-}NH}{CH} \qquad (1.5a)$$

Worse still it can, like the azide, undergo aminolysis by the amino component,

$$X{-}NH{-}CH(R_2){-}N{=}C{=}O + H_2N{-}CH(R_1){-}CO{-}Y \rightleftharpoons$$

$$\underset{\text{urea derivative}}{X{-}NH{-}CH(R_2){-}NH{-}CO{-}NH{-}CH(R_1){-}CO{-}Y} \qquad (1.6)$$

but the product is a urea derivative rather than a peptide; this derivative represents a failure in the synthesis. Worst of all, the urea is likely to be sufficiently similar to the peptide in its physical properties to make the two molecules very hard to separate. Fortunately, some hope is offered by the discovery that the urea can be cleaved in strong acid (Inouye and Watanabe, 1977).

Finally, the presence of HNO_2 in the activation mixture is a danger to the side-chains of tryptophan.

The yield of the azide coupling rarely exceeds 50% of the theoretical, but in spite of all the difficulties, when used with care, the method can give useful results.

The carbodiimide method (Sheehan and Hess, 1955) revolutionized peptide synthesis because of its simplicity and efficiency.

$$\mathrm{X{-}NH{-}\overset{\displaystyle R_2}{\overset{|}{C}}H{-}COOH} + \mathrm{H_2N{-}\overset{\displaystyle R_1}{\overset{|}{C}}H{-}CO{-}Y} + \mathrm{C_6H_{11}{-}N{=}C{=}N{-}C_6H_{11}} \rightleftharpoons$$

$$\mathrm{X{-}NH{-}\overset{\displaystyle R_2}{\overset{|}{C}}H{-}CO{-}NH{-}\overset{\displaystyle R_1}{\overset{|}{C}}H{-}CO{-}Y} + \mathrm{C_6H_{11}{-}NH{-}CO{-}NH{-}C_6H_{11}} \qquad (1.7)$$

The coupling proceeds via an *O*-acyl isourea.

$$\mathrm{X{-}NH{-}\overset{\displaystyle R_2}{\overset{|}{C}}H{-}COOH} + \mathrm{C_6H_{11}{-}N{=}C{=}N{-}C_6H_{11}} \rightleftharpoons \begin{array}{c}\mathrm{X{-}NH{-}\overset{\displaystyle R_2}{\overset{|}{C}}H{-}CO}\\ |\\ \mathrm{O}\\ |\\ \mathrm{C_6H_{11}{-}N{=}C{-}NH{-}C_6H_{11}}\end{array}$$

O-acyl isourea (1.7a)

$$\begin{array}{c}\mathrm{X{-}NH{-}\overset{\displaystyle R_2}{\overset{|}{C}}H{-}CO}\\ |\\ \mathrm{O}\\ |\\ \mathrm{C_6H_{11}{-}N{=}C{-}NH{-}C_6H_{11}}\end{array} + \mathrm{H_2N{-}\overset{\displaystyle R_1}{\overset{|}{C}}H{-}CO{-}Y} \rightleftharpoons$$

$$\mathrm{X{-}NH{-}\overset{\displaystyle R_2}{\overset{|}{C}}H{-}CO{-}NH{-}\overset{\displaystyle R_1}{\overset{|}{C}}H{-}CO{-}Y} + \mathrm{C_6H_{11}{-}NH{-}CO{-}NH{-}C_6H_{11}} \qquad (1.7b)$$

If the *O*-acyl isourea does not rapidly meet a molecule of amino component it decomposes to an *N*-acyl urea, which represents a loss of yield and is often tedious to remove.

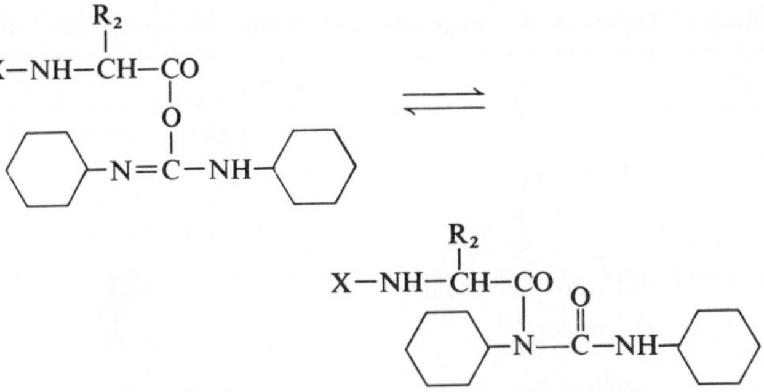

$$\text{X-NH-CH(R}_2\text{)-CO-O-C(=N-C}_6\text{H}_{11}\text{)-NH-C}_6\text{H}_{11} \rightleftharpoons \text{X-NH-CH(R}_2\text{)-CO-N(C}_6\text{H}_{11}\text{)-C(=O)-NH-C}_6\text{H}_{11}$$

N-acyl urea (1.8)

Racemization can be a serious problem when dehydrating agents such as the carbodiimides are used. One mechanism of racemization can occur when X— is an acyl group, and thus peptides themselves are among the compounds at risk. It proceeds via the intermediate formation of an oxazolone.

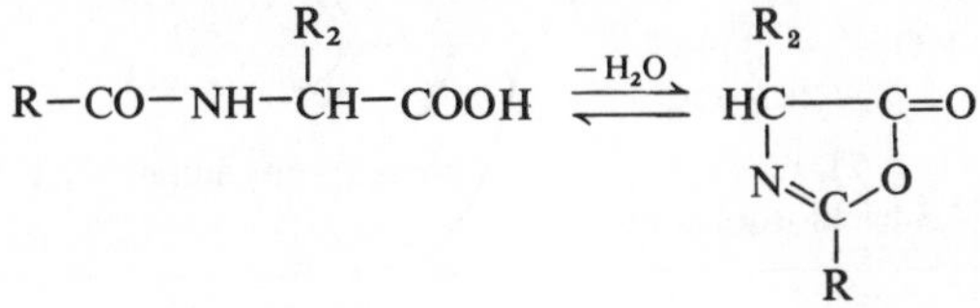

$$\text{R-CO-NH-CH(R}_2\text{)-COOH} \underset{}{\overset{-H_2O}{\rightleftharpoons}} \text{oxazolone (ring: HC(R}_2\text{)-C(=O)-O-C(R)=N-)}$$

oxazolone (1.9)

Although the oxazolone can react with an amino component to form a peptide bond

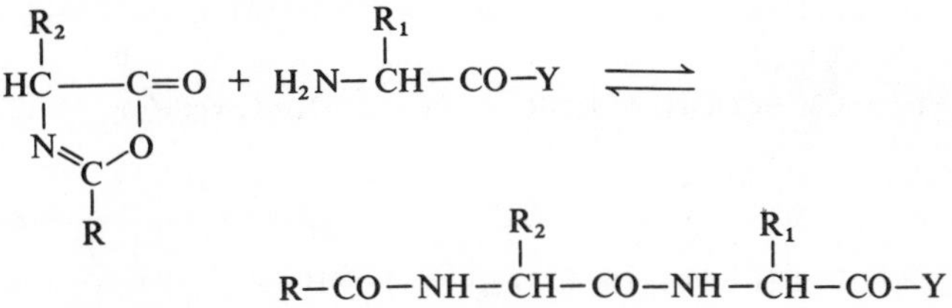

$$\text{oxazolone} + \text{H}_2\text{N-CH(R}_1\text{)-CO-Y} \rightleftharpoons \text{R-CO-NH-CH(R}_2\text{)-CO-NH-CH(R}_1\text{)-CO-Y} \quad (1.9a)$$

it can also tautomerize to forms in which the α-carbon is planar.

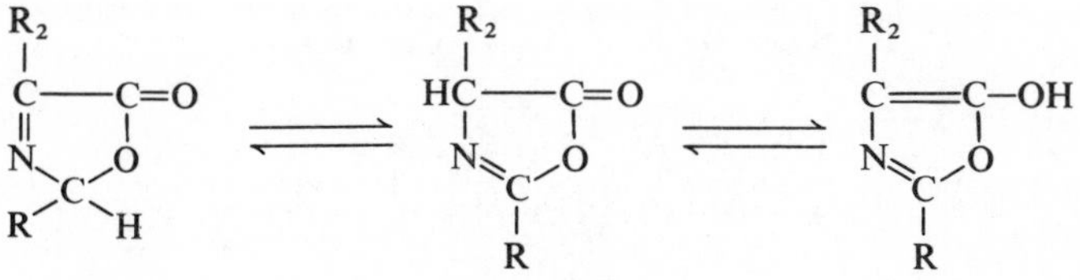

$$\text{C(R}_2\text{)=N-CH(R)-O-C(=O)} \rightleftharpoons \text{HC(R}_2\text{)-C(=O)-O-C(R)=N} \rightleftharpoons \text{C(R}_2\text{)=C(OH)-O-C(R)=N} \quad (1.9b)$$

Unless Reaction (1.9a) is much faster than Reaction (1.9b) the product of coupling will contain some racemized material.

Both of the difficulties just mentioned are lessened if the activated carboxyl component can react at once. For instance, it can be allowed to form an ester with certain alcohols, either *in situ* during the coupling or by prior reaction in the presence of the carbodiimide and the alcohol alone. The alcohols are chosen

Table 1.1. The alcohol components of some common active esters.

N-hydroxysuccinimide	*p*-nitrophenol
1-hydroxybenzotriazole	2-hydroxypyridine
2,4,5-trichlorophenol	2-mercaptopyridine

to have properties that promote rapid esterification and, subsequently, rapid and efficient aminolysis of the esters. Table 1.1 lists a few that have come into general use. In all cases, if the alcohol moiety is represented by R the esterification and aminolysis can be written as

$$X{-}NH{-}\overset{\overset{\displaystyle R_2}{|}}{CH}{-}COOH + ROH + C_6H_{11}{-}N{=}C{=}N{-}C_6H_{11} \rightleftharpoons$$

$$X{-}NH{-}\overset{\overset{\displaystyle R_2}{|}}{CH}{-}CO{-}O{-}R + C_6H_{11}{-}NH{-}CO{-}NH{-}C_6H_{11} \qquad (1.10)$$

$$X{-}NH{-}\overset{\overset{\displaystyle R_2}{|}}{CH}{-}CO{-}O{-}R + H_2N{-}\overset{\overset{\displaystyle R_1}{|}}{CH}{-}CO{-}Y \rightleftharpoons$$

$$X{-}NH{-}\overset{\overset{\displaystyle R_2}{|}}{CH}{-}CO{-}NH{-}\overset{\overset{\displaystyle R_1}{|}}{CH}{-}CO{-}Y + ROH \qquad (1.10a)$$

Some active esters decompose rapidly and must be used at once, others (the hydroxysuccinimido and *p*-nitrophenyl ones, for instance) can be stored, although racemization sets in if X— is an acyl group, as a result of electron withdrawal by the carbonyl group. (This tendency is suppressed if X— is a urethane group; we shall see later how this fact is exploited.) The dicyclohexylurea, formed when the reagent abstracts the elements of water in promoting the coupling, can be difficult to remove from the product. Other carbodiimides,

Table 1.2. Dicyclohexylcarbodiimide and two of its water-soluble analogues.

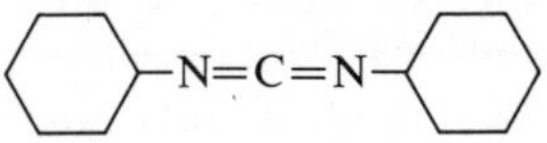

$$C_6H_{11}-N{=}C{=}N-C_6H_{11}$$

N,*N'*-dicyclohexylcarbodiimide

$$CH_3-CH_2-N{=}C{=}N-CH_2-CH_2-CH_2-N(CH_3)_2$$

N-ethyl-*N'*-(3-dimethylaminopropyl) carbodiimide

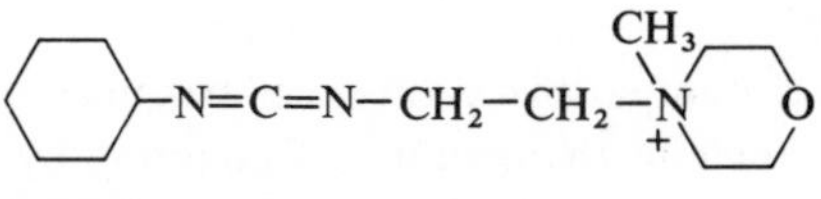

$$C_6H_{11}-N{=}C{=}N-CH_2-CH_2-\overset{+}{N}(CH_3)(CH_2CH_2)_2O$$

N-cyclohexyl-*N'*-(2-morpholinyl-(4)-ethyl) carbodiimide*

* Usually employed as the *p*-toluene sulphonate.

in which the substituent has a polar group (Table 1.2), give rise to ureas that are less like the wanted product in properties and which can be separated. Also, the polar-substituted carbodiimides are soluble in water and, since they react much more slowly with water than in the manner shown in Reaction (1.7), they can be used to promote couplings in aqueous solution.

There are many other methods, some of which, like the mixed-anhydride method (Wieland and Sehring, 1958; Anderson *et al.*, 1967) or the Woodward method (Woodward *et al.*, 1961), are quite popular. However, the ones discussed in detail are represented in the majority of syntheses and are also those most often employed so far in semisynthesis. Enzyme-catalysed coupling (e.g. Isowa *et al.*, 1977; Morihara and Oka, 1977) has been proposed for conventional synthesis, but has not yet been at all intensively exploited. It has found more application in semisynthesis (see Chapter 8), although so far only in special cases. Ways may exist of making the approach more generally applicable (p. 123).

Protection

Reversible and compatible

Apart from the question of the actual mechanism of the condensation, Reaction (1.1) indicates at once that many other products are possible unless the —NH_2 group of the carboxyl component and the —COOH group of the amino component are protected. We can for the moment take X— and Y— in the preceding Reactions to indicate the protecting groups.

The very simplest group will not always do—for example, the acetyl group would be easy to apply to the —NH_2 but the resulting amide bond would be chemically almost indistinguishable from the peptide bond. The product could not normally be deprotected without extensive decomposition. Protecting groups must be *reversible*—that is removed under conditions that do not damage the product—and fortunately many excellent ones have been developed. We shall see that many are removed as a result of rearrangements following protonation in strongly acidic media. Other favoured methods of deprotection involve exposure to basic or to reducing media. Whatever the conditions they should not be so severe as to break peptide bonds or damage side-chains.

The groups X— and Y— should be so chosen so as to make it possible to remove X without removing Y and *vice versa*. Protection is then said to be *compatible*. X— and Y— should be labile to quite dissimilar sets of conditions or, if they are both of the acid-labile type, then one of them must be susceptible to much more weakly acidic conditions than the other.

The need for differential, compatible protection becomes clear if we consider what must be done to extend the product of coupling by adding another amino-acid derivative to it. We must deprotect one of the two α-groups and repeat the coupling operation with the new derivative. For instance, suppose that we wish to add a residue to the product of Reaction (1.1). We can write the process formally as

$$\text{X—NH—}\underset{}{\overset{\text{R}_3}{\overset{|}{\text{CH}}}}\text{—COOH} + \text{H}_2\text{N—}\overset{\text{R}_2}{\overset{|}{\text{CH}}}\text{—CO—NH—}\overset{\text{R}_1}{\overset{|}{\text{CH}}}\text{—CO—Y} \rightleftharpoons$$

$$\text{X—NH—}\overset{\text{R}_3}{\overset{|}{\text{CH}}}\text{—CO—NH—}\overset{\text{R}_2}{\overset{|}{\text{CH}}}\text{—CO—NH—}\overset{\text{R}_1}{\overset{|}{\text{CH}}}\text{—CO—Y} + \text{H}_2\text{O} \qquad (1.11)$$

Unless we are using a method of coupling that involves prior activation of the carboxyl component, we must keep the —COOH group of the amino component protected, or it too will be activated.

Differential protection of side-chains

It was assumed in what was said above that the side-chain groups R_1, R_2, and R_3 were not themselves reactive and in need of protection. If, for example, R_1 were to be —CH_2CH_2COOH (the side-chain of glutamic acid) and R_2 were to be —$CH_2CH_2CH_2CH_2NH_2$ (the side-chain of lysine) this assumption would be invalid and these groups too must be protected. In general, it is best that the side-chain protection should be *differential* from that of the α-groups. The meaning of this term should be made clear if we consider that in Reaction (1.11), for instance, the —NH_2 protection of R_2 would have to have withstood the conditions that freed the α-NH_2 for coupling, or a peptide-like bond will be formed on the side-chain as well. As written there, the process does not call for differential protection of the —COOH groups, but if they were to be identically protected, molecules containing R_1 could never be used as a carboxyl component.

Minimal and total protection

Other side-chains are protected besides those with —NH_2 and —COOH groups. The —CH_2SH group of cysteine is so readily acylated, and so reactive in other ways, that it must be protected. The imidazole side-chain of histidine can be acylated under severe conditions and is sometimes protected. We shall see below that other side-chains that are sometimes protected are the indole group of tryptophan; the guanido group of arginine; the —OH of serine, threonine and tyrosine; the —$CONH_2$ of asparagine and glutamine; and the —SCH_3 group of methionine.

A synthesis in which as few types of side-chain are protected as possible is said to employ *minimal*, as opposed to *total*, protection. Syntheses employing minimal protection, total protection and virtually every intermediate compromise have been carried out with success. Questions of solubility often decide which approach is to be used, since the groups employed in conventional synthesis normally have the property of increasing the hydrophobic nature of the molecules to which they are attached. The effect is particularly marked if the group being protected is a polar one, and also depends on the extent of the protecting group's own hydrophobicity. Protected amino acids and peptides are usually totally insoluble in water, and organic solvents are used. However, the dipolar nature of the remaining —NH— and —CO— groups tends to prevent solution in solvents of really low polarity such as benzene and chloroform. Solvents of low polarity are found in general useful only for protected molecules of shorter chain length. For molecules of any size, solvents such as dimethylformamide and the cyclic ethers represent the best compromise (dimethylsulphoxide would probably rank as the best general solvent but it is a little too chemically reactive to be used in every instance). Even so, some protected products, particularly protected peptides of intermediate length (very approximately in the region of 20 residues), are very difficult to dissolve in anything. Often the chemist is driven to protecting functional groups that he would otherwise prefer to leave untouched simply to enhance solubility in organic solvents. In conventional syntheses this reason is sometimes the only one for departing from a strategy based on minimal protection which would otherwise have been preferable. In spite of solubility problems, it has been found best to carry out all semisyntheses with minimal protection—we shall see (p. 29) that doing so sometimes confers wholly unforeseen and major benefits.

Protecting reagents

This section will mention only a fraction of the vast number of protecting groups that have been proposed. I have restricted myself to those few that are used extremely commonly in conventional synthesis and that have been used at least once in semisynthesis.

We shall see in Chapters 2 and 4 that additional groups have been introduced into semisynthesis which are not usually thought of for conventional synthesis.

In particular, there has been an attempt to meet the solubility problems posed by the minimal nature of the protection in semisynthesis by employing polar, but chemically unreactive, substituents as protecting groups.

Amino-group protection. The urethan series are excellent. They are often applied by the aminolysis of the appropriate azide or chloroformate.

$$\mathrm{R{-}O{-}CO{-}N_3 + H_2N{-}\underset{}{\overset{R_1}{\overset{|}{C}}H}{-}CO{-}Y \rightleftharpoons R{-}O{-}CO{-}NH{-}\overset{R_1}{\overset{|}{C}}H{-}CO{-}Y + HN_3} \quad (1.12)$$

$$\mathrm{R{-}O{-}CO{-}Cl + H_2N{-}\overset{R_1}{\overset{|}{C}}H{-}CO{-}Y \rightleftharpoons R{-}O{-}CO{-}NH{-}\overset{R_1}{\overset{|}{C}}H{-}CO{-}Y + HCl} \quad (1.12a)$$

The group is labile to acid: the cleavage can proceed via the formation of a carbonium ion and an unstable carbamic-acid derivative.

$$\mathrm{R{-}O{-}CO{-}NH{-}\overset{R_1}{\overset{|}{C}}H{-}CO{-}Y + H^+ \rightleftharpoons \left[\underset{\text{carbamic-acid derivative (unstable)}}{HO{-}CO{-}NH{-}\overset{R_1}{\overset{|}{C}}H{-}CO{-}Y}\right] + \overset{+}{R}}$$

$$\rightleftharpoons \mathrm{CO_2 + H_2N{-}\overset{R_1}{\overset{|}{C}}H{-}CO{-}Y} \quad (1.13)$$

Where this mechanism controls the reaction (rather than the alternative, a bimolecular displacement) the more stable R^+ is, the weaker is the acid required for deprotection. The behaviour of three of the most popular urethan protecting groups is, as will be seen from Table 1.3, in line with what would be predicted on this basis. Their lability to acid is adequately differential for many purposes. (For more detail see Bláha and Rudinger, 1965 and Law, 1974.) For mechanistic reasons, they do not promote racemization of the adjacent α-carbon atom (cf. p. 8).

The Cbz- group can also be removed by reduction (Bergmann and Zervas, 1932)

$$\mathrm{C_6H_5{-}CH_2{-}O{-}CO{-}NH{-}\overset{R_1}{\overset{|}{C}}H{-}CO{-}Y + H_2 \xrightleftharpoons{\text{Pd/charcoal}} C_6H_5{-}CH_3 + CO_2 + H_2N{-}\overset{R_1}{\overset{|}{C}}H{-}CO{-}Y} \quad (1.14)$$

Table 1.3. Three urethan-type protecting groups.

Group*	Removed by
$C_6H_5-CH_2-O-CO-$ Cbz-	HBr/acetic acid
$CH_3-C(CH_3)_2-O-CO-$ Boc-	Trifluoroacetic acid
$C_6H_5-C_6H_4-C(CH_3)_2-O-CO-$ biphenylisopropyloxycarbonyl-	Dil. HCl

* In order of decreasing acid stability, from top to bottom.

giving a further degree of differential protection. The removal presumably proceeds, as before, via the carbamic acid. Unfortunately even the small quantity of sulphur found in a methionine- or cysteine-containing peptide tends to poison the catalyst. Most non-catalytic reducing agents, if they are strong enough to deprotect, bring about reductive cleavage of at least some peptide bonds.

The trifluoroacetyl group (Weygand and Csendes, 1952) is useful because it is stable to acid and labile to aqueous base and nucleophilic displacement. It can be applied either via the anhydride or in a reaction familiar to those who have worked on amino-acid sequences, via what is in effect an active ester.

$$CF_3-CO-S-CH_2-CH_3 + H_2N-\underset{\displaystyle R_1}{CH}-CO-Y \rightleftharpoons CF_3-CO-NH-\underset{\displaystyle R_1}{CH}-CO-Y + CH_3-CH_2-SH \qquad (1.15)$$

The removal,

$$CF_3-CO-NH-\underset{\displaystyle R_1}{CH}-CO-Y + OH^- \rightleftharpoons CF_3-COO^- + H_2N-\underset{\displaystyle R_1}{CH}-CO-Y \qquad (1.16)$$

proceeds reasonably rapidly at pH 11, which, though a little higher than one would wish, is within the range of pH that many proteins and peptides can survive.

The phthaloyl group (Kidd and King, 1948) is stable both to acid and to all but the strongest basic treatment. It can be efficiently applied under mild conditions by reaction with *N*-carboethoxyphthalimide (Nefkens *et al.*, 1960).

$$C_6H_4(CO)_2N{-}CO{-}O{-}CH_2{-}CH_3 + H_2N{-}\underset{\displaystyle |}{\overset{R_1}{C}}H{-}CO{-}Y \rightleftharpoons$$

$$C_6H_4(CO)_2N{-}\overset{R_1}{C}H{-}CO{-}Y + H_2N{-}CO{-}O{-}CH_2{-}CH_3 \qquad (1.17)$$

It can be removed with hydrazine (Radenhousen, 1895).

$$C_6H_4(CO)_2N{-}\overset{R_1}{C}H{-}CO{-}Y + H_2N{-}NH_2 \rightleftharpoons$$

$$C_6H_4(CO)_2(NH{-}NH) + H_2N{-}\overset{R_1}{C}H{-}CO{-}Y \qquad (1.18)$$

If care is taken this treatment can be made sufficiently mild to avoid damage to the product (Kappeler, 1963). As with the trifluoroacetyl group, its resistance to more normal means of deprotection, particularly its stability towards acid, makes it attractive for use in partnership with the urethan type of group. Its lability to hydrazine prohibits its use in routes that involve azide couplings.

Carboxyl protection. Protection is normally by esterification and all the usual methods of esterification have been employed. That used for the application of the t-butyl group is mildly unusual where biological materials are involved.

$$CH_2{=}C(CH_3)_2 + X{-}NH{-}\overset{R_2}{C}H{-}COOH \xrightleftharpoons{\text{Conc. } H_2SO_4}$$

$$X{-}NH{-}\overset{R_2}{C}H{-}CO{-}O{-}C(CH_3)_3 \qquad (1.19)$$

Yields are good but not quantitative even after reaction overnight. The prolonged exposure to concentrated H_2SO_4 does not harm most amino acids.

In contrast to the active esters used in coupling, one wants for protection a group that will remain as inert as possible until the time has come for deprotection. Table 1.4 gives a few commonly used groups.

Table 1.4. Some carboxyl-protecting groups.

Group	Effect of acid	Effect of reduction
$-O-CH_2-C_5H_4N$ 4-picolyl-	Removed only with difficulty	Removed
$-O-CH_2-C_6H_5$ benzyl-	Removed by HBr/acetic acid	Removed
$-O-C(CH_3)_3$ t-butyl-	Removed by trifluoroacetic acid	Stable
$-O-CH_3$ methyl-	Stable	Stable

The degree of acid lability of the esters $XNHCH(R_2)COOR$ once again roughly agrees with what is known of the stability of R^+. The benzyl ester appears to be slightly more stable to acid than the benzyloxycarboxyl amide and catalytic reduction is therefore all the more to be preferred over acid treatment where it can be carried out. It is practically obligatory to use reduction to remove the 4-picolyl group (Camble *et al.*, 1967). The t-butyl ester is extremely popular, but its use is not compatible with that of the Boc- group, since both are labile to trifluoroacetic acid. The t-butyl ester resists hydrazinolysis for steric reasons and its use is therefore compatible with coupling by the azide route (Reactions (1.2)–(1.4)).

The —CH_3 group, which has the advantage of being stable to the conditions that remove the others, was used early in the history of peptide synthesis (see Schröder and Lübke, 1966). It has not been much used of late, partly because of the rather high pH required for saponification and, in spite of reports that the presence of Cu^{2+} can prevent it, because of the danger of $\alpha \rightarrow \omega$ transpeptidation on deprotection (Reaction (1.20), overleaf.)

—SH protection. Really satisfactory protecting groups, that is ones that are removable under conditions mild enough to avoid damage to the product, have only relatively recently been developed. Table 1.5 indicates some of them.

Imidazole protection. Table 1.6 indicates some of the groups available if it is decided to protect the reactive nitrogen of the histidine side-chain. Urethan protecting groups are, when on the imidazoyl nitrogen, differentially labile to those on the amino nitrogen (Elliott and Morris, 1960). The reagents based on 2,2,2-trifluoroethyl-1-amine have the stability of their urethan moiety. They can be applied by reaction with the trifluoroacetyl ester of the appropriate alcohol.

Guanido protection. This group is sometimes protected by nitration. This has, as usual in protection, not only the effect of suppressing the chemical reactivity

$$\text{X–NH–CH(R}_3\text{)–CO–NH–CH(CH}_2\text{–CH}_2\text{–CO–O–CH}_3\text{)–CO–NH–CH(R}_1\text{)–CO–Y} + \text{OH}^- \rightleftharpoons \underset{\text{desired product}}{\text{X–NH–CH(R}_3\text{)–CO–NH–CH(CH}_2\text{–CH}_2\text{–COO}^-\text{)–CO–NH–CH(R}_1\text{)–CO–Y}} + \text{CH}_3\text{OH}$$

$$\Bigg\downarrow\!\uparrow$$

$$\underset{\text{imide intermediate}}{\left[\text{X–NH–CH(R}_3\text{)–CO–NH–}\overset{\text{CH}_2\text{–CH}_2\text{–CO}}{\underset{\text{CH}\text{——CO}}{|}}\!\!>\text{N–CH(R}_1\text{)–CO–Y}\right]} \rightleftharpoons \underset{\text{rearranged product}}{\text{X–NH–CH(R}_3\text{)–CO–NH–CH(COO}^-\text{)–CH}_2\text{–CH}_2\text{–CO–NH–CH(R}_1\text{)–CO–Y}} \qquad (1.20)$$

Table 1.5. Some thiol-protecting groups.

Group	Removed by
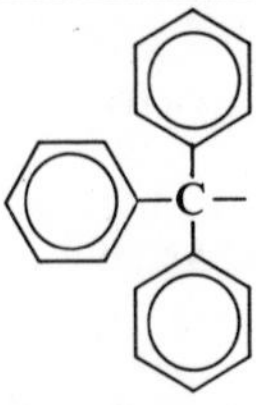 C− (three phenyl rings) trityl-	$HgCl_2$; trifluoroacetic acid; I_2
$CH_3-C(CH_3)_2-S-$ t-butylmercapto-	Thiols
$CH_3-CO-NH-CH_3-$ acetamidomethyl-	Mercuric acetate; I_2

Table 1.6. Some imidazole-protecting groups.

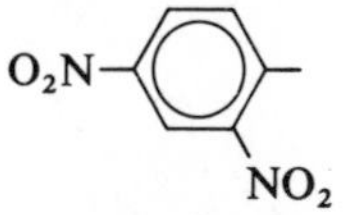 O_2N-(ring)$-$, NO_2 2,4-dinitrophenyl-	Stable to acid, removed by thiols
$R-O-CO-NH-CH(CF_3)-$ Urethan derivatives of 2,2,2-trifluoroethyl-1-amine	Has the usual acid stability of the urethan moiety
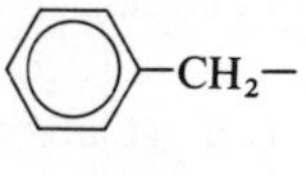 (ring)$-CH_2-$ benzyl-	Stable to acid, removed (with difficulty) by reduction

of the group, but also of reducing the polarity of the molecule as a whole. Since the guanido group is very strongly basic it is, even when unmodified, effectively protected from side-reactions by protonation under most of the conditions encountered in peptide synthesis. The desire to suppress its polarity and enhance solubility in organic solvents rather than to avoid side-reactions has probably been decisive in a large number of the instances in which the choice was taken to protect it.

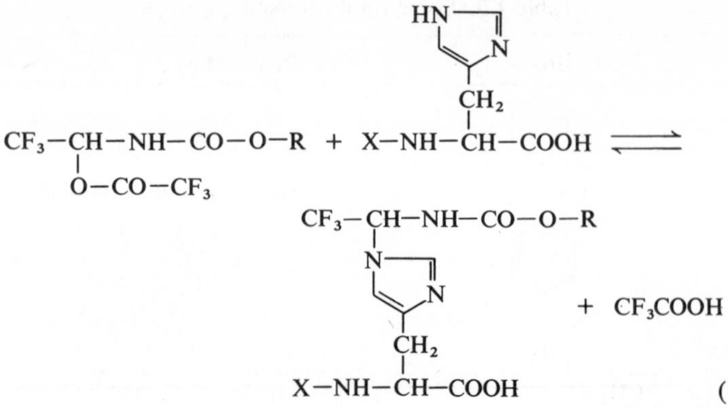

$$CF_3{-}\underset{\displaystyle O{-}CO{-}CF_3}{CH}{-}NH{-}CO{-}O{-}R \;+\; X{-}NH{-}\overset{\displaystyle \text{HN—(imidazole)—N} \atop \displaystyle CH_2}{CH}{-}COOH \;\rightleftharpoons$$

$$CF_3{-}\underset{\displaystyle \text{N—(imidazole)—N} \atop {\displaystyle CH_2 \atop \displaystyle X{-}NH{-}CH{-}COOH}}{CH}{-}NH{-}CO{-}O{-}R \;+\; CF_3COOH \qquad (1.21)$$

Other side-chains. The —OH groups of serine, threonine, and tyrosine are in theory capable of being acylated during coupling. They can be protected by etherification, but, as before, it does not seem to be essential to do so under most conditions unless, once again, one wishes to influence solubility. Indeed the protection of the —OH of tyrosine may do more harm than good because of intramolecular attack by the protonated leaving group on deprotection (Iselin, 1962).

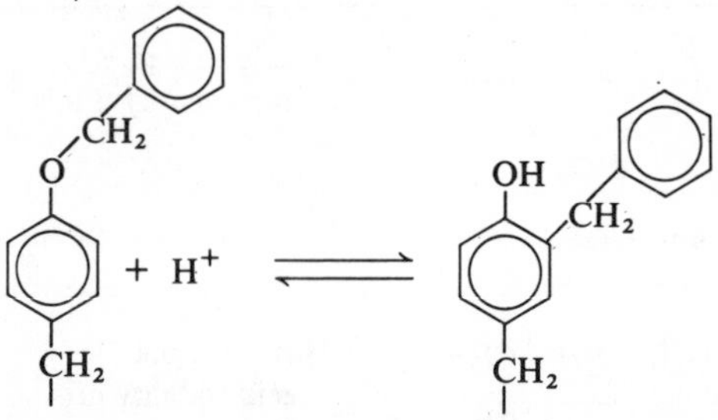

$$C_6H_5{-}CH_2{-}O{-}C_6H_4{-}CH_2{-} \;+\; H^+ \;\rightleftharpoons\; HO{-}C_6H_3(CH_2{-}C_6H_5){-}CH_2{-} \qquad (1.22)$$

As the attack is *intra*molecular the use of a scavenger (p. 146) could very possibly be ineffective.

Attempts are sometimes made to protect the indole ring of tryptophan, mainly from oxidative degradation, and something will be said of this in Chapter 4. The protection of amide side-chains against dehydration (e.g. Akabori *et al.*, 1961) and the thioether side-chain of methionine against alkylation (in particular by R^+ formed on deprotection of other groups) is occasionally undertaken but many successful syntheses attest to the fact that it is not essential.

Dangers on deprotection

Even if the synthesis has gone well, all can quite easily be lost as a result of damage on deprotection. In particular, in view of the considerable preponderance of acid-labile groups, the special problems involved in removing them are of great importance. Many proteins and peptides turn out to be surprisingly resistant to exposure to even the strongest acids, so long as water is completely absent and hydrolysis is thereby prevented. If this precaution is neglected, there is progressive cleavage of side-chain amide and peptide bonds, of which the former are in general the more susceptible. Damage is further minimized by using strong acids only for the briefest possible interval, and at as low a temperature as possible. Anhydrous trifluoroacetic acid has proved extremely valuable. It is readily removed by evaporation and the range of groups that can be deprotected can be extended by dissolving HBr in it (e.g. Guttman and Boissonas, 1958). The range can be reduced by diluting it in an organic solvent, particularly one of low polarity (e.g. Stelakatos and Argryoupoulos, 1970). Anhydrous HF probably represents the most severe acidic reagent in general use: proteins survive exposure to it markedly better than do the tissues of the operator.

Acid-promoted degradation of the side-chain of tryptophan is a more serious problem than hydrolysis. It has been the cause of the abandonment of at least one synthesis at an advanced stage (Sano and Kurihara, 1969). The problem with tryptophan is aggravated, and extended to other side-chains, if, as is so often true, the residue of the protecting group after cleavage is a free carbonium ion. These ions can alkylate other side-chains, notably that of tryptophan (Alakhov *et al.*, 1970) either at once or as the acid is removed after deprotection. As mentioned on p. 146, the addition of competing agents ('scavengers') such as $C_6H_5OCH_3$ and $CH_3CH_2SCH_3$ decreases, but does not always abolish, the damage from this source. Thus strong acids can do more damage on deprotection than a trial experiment on a sample of native protein might suggest.

Finally, strongly acidic media promote the transfer of the —COOH group forming a peptide bond with the —NH_2 group of serine or threonine to the side-chain —OH. Fortunately, this N→O acyl shift is reversible on brief exposure to mild aqueous base (p. 141).

The strategy of peptide synthesis

Stepwise and fragment-condensation syntheses

The various coupling and protection reagents can be deployed in practice in either of two ways. In *stepwise* syntheses one starts with the terminal residue and adds the others in the correct order, one by one, until the molecule is finished. Since almost all coupling methods involve the activation of the —COOH group of the carboxyl component, it is most sensible to begin with the carboxyl-terminal residue of the desired sequence and extend it in the direction of the amino terminus as was done in Reaction (1.11)—that is, in the opposite direction to that in biosynthesis. A single application of the two types of operation, coupling and then deprotection of the new terminal —NH_2 group, constitutes one cycle of the stepwise process.

The stepwise approach can be contrasted with the alternative, *fragment-condensation* syntheses in which separate portions of the chain are assembled in a series of independent stepwise syntheses. These portions are then combined, either stepwise, in their proper order starting from one end, or more usually into larger subfragments which are in their turn combined into still larger ones until all are joined in the final molecule. We shall see that semisynthetic versions exist of both the stepwise and the fragment-condensation strategies.

Solid-phase stepwise synthesis: automation

Merrifield proposed a method (reviewed by Stewart and Young, 1969) in which the carboxyl-terminal amino acid is first reversibly attached, by a covalent bond, to a macroscopic solid support. The stepwise synthesis proceeds in the normal way, but at each stage excess reagent, by-products, and solvents can be completely washed away without losing the wanted products. Yields are thereby much improved over those in syntheses in free solution where extraction, crystallization, or chromatography are used to isolate the growing peptide chain after each step. Since the growing polypeptide chain is not purified at each cycle it is both more necessary and easier to use a large excess of carboxyl component to drive the coupling to completion. Furthermore, the process lends itself readily to automation. It is quite simple to design a piece of equipment that will, in the right order and at the proper intervals, deliver and then remove the active esters and other reagents that are required for the complete synthesis.

Failure sequences and other problems

Is then the automated solid-phase method the answer to all the problems of peptide synthesis? What is the point in considering the fragment-condensation method as an alternative strategy of conventional synthesis—or for that matter semisynthesis—at all? The answer to such queries—which were never raised by those that developed the automated methods—lies in the computation of the overall yield of a multistep process. If we synthesize a peptide of n residues with an average efficiency per step of $q\%$ then the overall yield is $(q/100)^{n-1} \times 100\%$. Table 1.7 gives some numerical results obtained with this expression. Even for very small proteins, where n is in the region of 100, q must be extremely high for the product to contain any appreciable amount at all of the correct structure. The problem is not merely one of low yield of the wanted product, but its contamination by molecules that have one or more residues deleted. The failure of one coupling step does not necessarily mean that the growth of the peptide chain is terminated at that point, and all the other residues could well be added on in the subsequent steps of the synthesis. Molecules of incorrect structure that arise in this way are called *failure sequences*. Apart from the fact that they will predominate in a product of any but the shortest length unless the coupling efficiency is extremely high, they will be a heterogeneous mixture.

Table 1.7. Stepwise synthesis. The percentage of the final product that has the correct sequence, as a function of chain length and efficiency per cycle.

Number of residues (n)	Average percentage efficiency per cycle (q)			
	80	90	95	98
3	64	81	90	96
10	13	39	63	83
30	1.5×10^{-1}	4.7	23	56
100	2.5×10^{-8}	2.6×10^{-3}	6.2×10^{-1}	13
300	1.1×10^{-27}	2.1×10^{-12}	2.2×10^{-5}	2.4×10^{-1}

All these factors, together with the fact that the lack of one amino acid among many will often have only a slight effect on physical properties, tend to make it quite difficult to isolate the wanted product from the mixture. Something can be done if the deletions have a severe effect on the folding of the chain into its proper three-dimensional structure but it is still likely that even after all possible treatments the products would remain heterogeneous. The extent of biological activity is no certain guide to relative purity, since studies on mutant proteins that have natural deletions show that although deletions can often inactivate, they by no means always do so.

There is also a large cumulative exposure to the reagent that deprotects the α-NH_2 of each newly inserted residue, in particular of those parts of the chain that are made early on. Quite apart from any damage to the main chain that this might do, side-chain protecting groups that are compatible with the N^{α} protection when only a few minutes total exposure are involved may begin to be lost after a total exposure of a day or more (e.g. Chillemi and Merrifield, 1969). If the side-chain group thus unmasked can be acylated, it can serve as a growing point for a new polypeptide chain.

At present all these problems probably weigh decisively against the use of stepwise synthesis for long polypeptides. The present drawbacks should not prevent us from looking forward to a time when technical improvements make it as reliable for long chains as it probably now is for short ones. Even if that time were never to come, there are likely to be some proteins needed for practical purposes rather than for research for which an unhomogeneous sample might be acceptable in the absence of a better alternative source.

Fragment condensation: the virtues of intermediate purification

We have to form the same number of peptide bonds even if we turn to the use of fragment condensation, and of course the fragments themselves have to be made by stepwise synthesis. Does not Table 1.6 apply equally well? It does in a sense, but at least if we pause when we have made our fragments and seek to purify out the molecules that are damaged, have deletions, or peptides growing from the side-chains, the heterogeneity of our product is reduced even if the yield is not increased. It is, in any case, usually a great deal easier to separate, say, a 9-residue failure sequence from the 10-residue wanted product than if the chain lengths were 99 and 100—the proportional effect of the deletion on the properties of the molecules will be so much smaller in the large molecules. Also, we can minimize the disadvantage stemming from the $(q/100)^{n-1} \times 100$ law. For instance the building up of, say, the 61st to the 70th residue in a stepwise synthesis can begin on no greater quantity than the yield at the 60th residue. If we make these 10 residues as a separate fragment, we can decide for ourselves how much material to begin with.

Of course, the coupling of the fragments involves a reaction between two bulky molecules rather than, as in stepwise synthesis, between one small one and a bulky one. Steric hindrance and the difficulty of obtaining adequately high concentrations of reactants become even more serious problems, and many fragment-condensation syntheses have had quite dismal yields at late stages. (We shall see later that the practice of using minimal protection in semisynthesis, and the tendency to choose polar protecting groups, can occasionally lead to quite remarkable departures from the general rule that large fragments couple together less well than do small ones. When the components of the coupling are left with as much as possible of their original non-covalent bonding properties, the forces that stabilize the conformation of the native protein can sometimes steer the coupling rapidly to completion.)

It is probably true that those who favour fragment condensation over stepwise synthesis, in spite of the greater demands on their time and skill, do so as much because of the greater reliability of the structure of the product as because they hope for better yields. Some workers compromise by using manual or automated solid-phase synthesis to make their fragments.

SEMISYNTHESIS

The conventional synthesis of a protein of well-defined structure is, for the reasons given in the last few pages, rather difficult. We ought at least to consider what benefits there might be in exploiting the supply of ready-made peptides obtainable from natural sources.

Several proteins may be cleaved into a small number of fragments which recombine through non-covalent interactions to form complexes exhibiting physical, and in some cases biological, properties similar to those of the parent protein. Examples of such proteins include pancreatic ribonuclease (Richards, 1958; Richards and Vithayathil, 1959), *Staphylococcus aureus* nuclease (Taniuchi *et al.*, 1967), soyabean trypsin inhibitor (Kato and Tominaga, 1970), thioredoxin (Holmgren, 1972; Slaby and Holmgren, 1975), cytochrome *c* (Fisher *et al.*, 1973; Corradin and Harbury, 1974b; Hantgan and Taniuchi, 1976, 1977; Harris and Offord, 1977; Wilgus *et al.*, 1978), Staphylococcal enterotoxin C (Spero *et al.*, 1976), *Staphylococcus aureus* penicillinase (Carrey and Pain, 1977), myoglobin (Hagenmaier *et al.*, 1978), somatotropin (Li and Bewley,

1976), and immunoprotein fragments (e.g. Hochman *et al.*, 1976). Where this occurs, natural fragments can be combined with synthetic or modified peptides to form non-covalent semisynthetic analogues (e.g. for ribonuclease, Scoffone *et al.*, 1967; for *S. aureus* nuclease, Ontjes and Anfinsen, 1969; for cytochrome *c*, Harris and Offord, 1977; and for somatotropin, Li *et al.*, 1978). A semisynthetic sample of the fragment F_V of an immunoglobulin A molecule has been prepared by non-covalent association, but no sequence changes have yet been introduced. These operations are discussed in more detail in Chapter 8.

Alternatively, the structures of a few proteins permit the formation of covalent semisynthetic complexes, held together by disulphide bridges (e.g. for insulin, see Chapter 8).

Useful though these approaches are, it would usually be preferable to combine the natural and synthetic portions by means of normal peptide bonds. This more general approach to semisynthetic proteins can be dealt with in the terms of the discussion of conventional synthesis in the preceding few pages. For instance, suppose that we wished to synthesize by stepwise means a protein that differs from a natural one at a site fairly near one end of the molecule. We could in theory make the product by a few cycles of stepwise synthesis, using as a starting material the natural molecule after truncation by the appropriate number of cycles of some stepwise degradative technique. Alternatively, suppose that the site of the proposed modification were to be deep within the sequence, we could then use fragment condensation, and all but the fragment carrying the alteration should be obtainable by cleaving the naturally occurring molecule. Even the fragment with the alteration might be obtainable by modification of the natural product rather than by conventional synthesis.

What we have just done is to point out the possibility of *stepwise semisynthesis* and *fragment-condensation semisynthesis* in strict analogy to the conventional form of these techniques. There are several potential advantages in these approaches. First, the naturally obtained intermediates are of absolutely authentic sequence with little or no possibility of having suffered racemization. Second, the effort in obtaining them in reasonable quantities is, for many proteins, not large and (in sharp contrast to conventional syntheses) only slightly dependent on the chain length. Thirdly, the number of couplings is spectacularly reduced (see, for example, Fig. 3.1, p. 56).

The plan of this book

The remainder of this book describes the steps involved in both stepwise and fragment-condensation semisynthesis. The two approaches have, as the steps that distinguish them from conventional synthesis, the cleavage and separation of the fragments of the native molecule (Chapter 3) and, if possible, the modification rather than total synthesis of the fragment that is to carry the site of alteration (Chapter 5). They also involve, in common with conventional synthesis, protection (in part before cleavage (Chapter 2) and in part after it (Chapter 4)), coupling (Chapter 6), deprotection and final purification (Chapter

7). Although we are entitled to anticipate that the naturally obtained fragments would tend to be of higher quality than synthetic ones, they will not always escape damage during separation and handling, nor occasional failures in protection and coupling. It is therefore necessary, as with conventionally obtained materials, to characterize products and intermediates as far as possible (Appendix 1). The Appendix also collects together a few generally useful techniques. The products of semisyntheses so far published are compared with conventional ones in Chapter 8, which also discusses the problems that remain to be overcome and the future possibilities for the technique.

CHAPTER 2

Protection before cleavage

The central problem of semisynthesis: differential protection of side-chains

If we wish to couple two fragments obtained by conventional synthesis, they will have their side-chain protection already in position. This built-in protection does not occur in natural fragments and must be supplied. Consider, for instance, the coupling between glutamyl-lysyl-arginine and phenylalanyl-lysyl-aspartic acid (Reaction (2.1)). The carboxyl component will require protection on those of its side-chains that are liable to acylation as well as on its

$$\mathrm{H_2N{-}CH(CH_2CH_2COOH){-}CO{-}NH{-}CH(CH_2CH_2CH_2CH_2NH_2){-}CO{-}NH{-}CH(CH_2CH_2CH_2NH{-}C({=}NH)NH_2){-}COOH} \; + \; \mathrm{H_2N{-}CH(CH_2C_6H_5){-}CO{-}NH{-}CH(CH_2CH_2CH_2CH_2NH_2){-}CO{-}NH{-}CH(CH_2COOH){-}COOH} \longrightarrow$$

$$\mathrm{H_2N{-}CH(CH_2CH_2COOH){-}CO{-}NH{-}CH(CH_2CH_2CH_2CH_2NH_2){-}CO{-}NH{-}CH(CH_2CH_2CH_2NH{-}C({=}NH)NH_2){-}CO{-}NH{-}CH(CH_2C_6H_5){-}CO{-}NH{-}CH(CH_2CH_2CH_2CH_2NH_2){-}CO{-}NH{-}CH(CH_2COOH){-}COOH} \qquad (2.1)$$

side-chain —COOH groups and its α-NH_2 group. The α-COOH group is left free. If the product is ever to be extended by coupling at the amino terminus the α-NH_2 will have to be protected by a group that can be removed without any others being affected.

The amino component in Reaction (2.1), similarly, will need its α-NH_2 free while its acylatable side-chains must be protected. If a carbodiimide is to be used for activation *in situ* (p. 6) the —COOH groups must also be protected or they might participate in coupling reactions. When —COOH protection is employed, the α-COOH must be differentially protected from the side-chain groups if it is ever to be possible to extend the product on from the carboxyl terminus.

It would be quite simple to protect the α- and side-chain —NH_2 groups with one type of reagent and the α- and side-chain —COOH groups with another. But we have seen that we need to protect α- and side-chain groups differentially. At first sight this requirement seems difficult to meet if our starting material is the unprotected peptide. The chemical reactivity of a group is not normally very much dependent on whether it is on the α-carbon or on the side-chain. How can we, say, treat a peptide with two —NH_2-protecting reagents and expect one to react exclusively with the α—NH_2 and one to react exclusively with the side-chain —NH_2? In conventional synthesis, by contrast, we can appeal to the ability of the —COOH and —NH_2, when on the same carbon atom, to form metal chelates

$$2\,R{-}CHNH_2{-}COO^- + Cu^{2+} \rightleftharpoons \begin{array}{c} R\diagdown\quad\quad\quad\quad\quad\diagup R \\ HC{-}NH_2\;\;H_2N{-}CH \\ |\quad\quad\cdot\cdot\,Cu\,\cdot\cdot\quad\quad| \\ C{-}O\diagup\quad\quad\diagdown O{-}C \\ /\!/\quad\quad\quad\quad\quad\quad\diagdown\!\diagdown \\ O\quad\quad\quad\quad\quad\quad\quad O \end{array} \tag{2.2}$$

and thus to be protected. The side-chain can be protected by any desired reagent, the metal removed, and the α-group differentially protected.

$$\text{Lysine} \xrightarrow{Cu^{2+}} \text{Cu chelate} \xrightarrow[\text{(Reaction (1.12))}]{(CH_3)_3C{-}O{-}CO{-}N_3} N^{\varepsilon}\text{-Boc-Lys chelate} \xrightarrow{H_2S} N^{\varepsilon}\text{-Boc-Lys} \xrightarrow[\text{(Reaction (1.12a))}]{C_6H_5{-}CH_2{-}O{-}CO{-}Cl} N^{\varepsilon}\text{-Boc-}N^{\alpha}\text{-Cbz-Lys} \tag{2.3}$$

Differential protection of the —COOH groups of amino acids can be achieved in a precisely analogous way.

Such procedures are difficult to apply to peptides and proteins. It was the resulting problem of differential protection that largely prevented semisynthesis from being introduced earlier than it was.

An answer: prior protection

The peptide bond is itself a differential protection of the α-COOH and —NH_2 groups. As Goldberger and Anfinsen pointed out (1962), reversible acylation of ε-NH_2 groups of a protein before cleavage will give rise to fragments that are, as far as the —NH_2 groups are concerned, half-way to being protected in precisely the way we require. Some other, compatible group can then be applied to the α-NH_2.

$$H_2N\text{—}[\text{chain with three side-chain }NH_2]\text{—}COOH$$

↓

$$R_1\text{—}HN\text{—}[\text{chain with three side-chain }R_1\text{—}NH]\text{—}COOH$$

↓

$$R_1\text{—}HN\text{—}[R_1\text{—}NH]\text{—}COOH + H_2N\text{—}[R_1\text{—}NH]\text{—}COOH + H_2N\text{—}[R_1\text{—}NH]\text{—}COOH$$

↓

$$R_1\text{—}HN\text{—}[R_1\text{—}NH]\text{—}COOH + R_2\text{—}HN\text{—}[R_1\text{—}NH]\text{—}COOH + R_2\text{—}HN\text{—}[R_1\text{—}NH]\text{—}COOH \quad (2.4)$$

We shall see later that, although the equivalent problem of —COOH protection can in theory be solved in the same way, other approaches (Backer and Offord, 1968; Offord, 1969a; 1972) are in general found to be necessary. It is advantageous to protect —SH groups before cleavage in order to facilitate separation of the fragments: the protection of the imidazole group, the only other side-chain normally included in the minimal protection schemes of semisynthesis, can be carried out either before or after cleavage.

This Chapter describes the application of prior protection to proteins.

Amino protection

All the reversible groups mentioned in Chapter 1—in particular Boc-, Cbz-, trifluoroacetyl, and phthaloyl—can be readily applied to proteins. (The trifluoroacetyl group was in fact the one proposed for semisynthesis by Goldberger and Anfinsen in 1962, although protein trifluoroacetylation has in the event been used more to limit tryptic cleavage for the purposes of amino-acid sequence determination.)

Examples (2.1), (2.2) and (2.3) indicate the methods for applying these groups that have been found satisfactory in this Laboratory.

We saw in Chapter 1 that many of the —NH_2-protecting groups are removed by acid and that of those that are acid-stable, most require conditions for removal that at the very least carry a risk of damaging the protein. In order to obtain differential protection that can be removed without damage, it is worthwhile to seek other acid-stable groups, capable of removal under safe conditions. The interest in safe acid-stable groups becomes even stronger when it is realized that the Edman reaction, which is otherwise the method of choice for use in stepwise semisynthesis (Chapter 3), involves the use of strong acid. To meet this need Busse and Carpenter (1974, 1976), Busse *et al.* (1974), Saunders (1974), and Saunders and Offord (1977a, b) have employed protecting groups which are in effect acyl methionines, applied to the protein as their active esters.

$$\begin{array}{l} \quad\quad\quad CH_3 \\ \quad\quad\quad | \\ \quad\quad\quad S \\ \quad\quad\quad | \\ \quad\quad\quad CH_2 \\ \quad\quad\quad | \\ \quad\quad\quad CH_2 \quad\quad\quad\quad\quad\quad\quad NH_2 \\ \quad\quad\quad | \quad\quad\quad\quad\quad\quad\quad\quad\quad | \\ R{-}NH{-}CH{-}CO{-}\text{ester} + H_2N\text{———}COOH \longrightarrow \end{array}$$

$$\begin{array}{l} \quad\quad\quad\quad\quad\quad\quad\quad\quad\quad\quad\quad\quad CH_3 \\ \quad\quad\quad CH_3 \quad\quad\quad\quad\quad\quad\quad | \\ \quad\quad\quad | \quad\quad\quad\quad\quad\quad\quad\quad\quad S \\ \quad\quad\quad S \quad\quad\quad\quad\quad\quad\quad\quad\quad | \\ \quad\quad\quad | \quad\quad\quad\quad\quad\quad\quad\quad\quad CH_2 \\ \quad\quad\quad CH_2 \quad\quad\quad\quad\quad\quad\quad | \\ \quad\quad\quad | \quad\quad\quad\quad\quad\quad\quad\quad\quad CH_2 \\ \quad\quad\quad CH_2 \quad\quad\quad\quad\quad\quad\quad | \\ \quad\quad\quad | \quad\quad\quad\quad\quad R{-}NH{-}CH{-}CO{-}NH \\ R{-}NH{-}CH{-}CO{-}NH\text{——————}|\text{——}COOH \end{array} \tag{2.5}$$

Busse and Carpenter used R = CH_3CO—; Saunders and Offord have in addition used R = Cbz-, ethyloxycarbonyl-, and phenylethyloxycarbonyl-. Example (2.4) gives experimental details for a group of this type.

Remarkably enough for groups that were designed to meet the need for safe deprotection the reagent proposed—which is probably the only possible one—is CNBr in acid solution.

$$\begin{array}{l} \quad\quad\quad\quad\quad\quad\quad\quad\quad\quad\quad\quad\quad CH_3 \\ \quad\quad\quad CH_3 \quad\quad\quad\quad\quad\quad\quad | \\ \quad\quad\quad | \quad\quad\quad\quad\quad\quad\quad\quad\quad S \\ \quad\quad\quad S \quad\quad\quad\quad\quad\quad\quad\quad\quad | \\ \quad\quad\quad | \quad\quad\quad\quad\quad\quad\quad\quad\quad CH_2 \\ \quad\quad\quad CH_2 \quad\quad\quad\quad\quad\quad\quad | \quad\quad\quad\quad\quad \xrightarrow{CNBr} \\ \quad\quad\quad | \quad\quad\quad\quad\quad\quad\quad\quad\quad CH_2 \\ \quad\quad\quad CH_2 \quad\quad\quad\quad\quad\quad\quad | \\ \quad\quad\quad | \quad\quad\quad\quad\quad R{-}NH{-}CH{-}CO{-}NH \\ R{-}NH{-}CH{-}CO{-}NH\text{——————}|\text{——}COOH \end{array}$$

$$\begin{array}{l} \quad\quad\quad\quad\quad\quad H_2 \\ \quad\quad\quad\quad\quad\quad C \\ \quad\quad\quad\quad\quad / \quad \backslash \\ \quad\quad\quad\quad H_2C \quad\quad O \quad\quad\quad\quad\quad\quad NH_2 \\ \quad\quad\quad\quad | \quad\quad\quad\quad | \quad\quad + \quad\quad\quad\quad | \\ 2\,R{-}NH{-}CH{-}C \quad\quad H_2N\text{———}COOH \\ \quad\quad\quad\quad\quad\quad\quad \| \\ \quad\quad\quad\quad\quad\quad\quad O \end{array} \tag{2.6}$$

This reaction is familiar in sequence determination as a means of cleaving proteins at methionine residues, and of course the acyl methionines can only be used in proteins that lack methionine altogether. If this condition is met (as it is in insulin) the deprotection is likely to do little damage. Nonetheless, that two sets of workers should have found it necessary to propose such a bizarre scheme of protection and deprotection is an indication of the relative shortage, even after 70 years of development of conventional synthesis, of suitable acid-stable —NH_2-protecting groups.

Geiger *et al.* (1975a) have proposed the use in semisynthesis of the methylsulphonylethyloxycarbonyl group (Msc-) of Tesser (1975). It is stable to acid, but labile to base. The basic conditions required are still a little stronger than one could wish, but Geiger has shown that insulin resists them, and it is reasonable to expect that some others will be found that will behave equally well. Note 6 to Example (2.8) describes the application of this group.

Amino protection by polar groups

Acetimidylation. Lysine in proteins can be N^{ϵ}-acetimidylated (Hunter and Ludwig, 1962).

$$\underset{\text{methyl acetimidate}}{CH_3-\overset{\overset{+}{N}H_2}{\overset{\|}{C}}-O-CH_3} + R-NH_2 \rightleftharpoons CH_3-\overset{\overset{+}{N}H_2}{\overset{\|}{C}}-NH-R + CH_3OH \qquad (2.7)$$

Example (2.5) gives experimental details, including the preparation of the reagent.

The acetimidyl amino group is still ionizable

$$CH_3-\overset{\overset{+}{N}H_2}{\overset{\|}{C}}-NH-R \rightleftharpoons CH_3-\overset{NH}{\overset{\|}{C}}-NH-R + H^+ \qquad (2.8)$$

and so the solubilizing effect of the positive charge of the lysine ($pK \simeq 10$) is not lost.

Although the chemical reactivity of the —NH_2 groups is satisfactorily suppressed, the retention of charge means that acetimidylated proteins are likely to suffer very little change in conformation from the native state. There is some increase in the bulk of the groups protected, but these are normally found on the outside of the protein, where the change in size can usually be accommodated without having to move other parts of the molecule. Indeed, as has already been mentioned, there are signs that if we employ protection that has such a minimal effect, then the coupling between large peptides, far from being much slower than that between small ones can on occasion actually be faster. This effect is the result of the bringing together of the α-COOH and α-NH_2 groups by the normal interplay of non-covalent forces that stabilize the tertiary structure of the intact protein.

Even biological activity can be retained on acetimidylation, as, for example, Nureddin and Inagami (1969) have shown for trypsin. Of course, when an —NH_2 group is very intimately concerned in the mechanism of action of a protein, even the relatively small changes brought about by acetimidylation can have a marked effect. For instance, the —NH_2 group of lysine-13 of the enzyme triosephosphate isomerase is believed to take an essential part in the interaction between enzyme and substrate (Banner *et al.*, 1975) and it is probably for this reason that the acetimidylated enzyme has very low catalytic activity (Milman, J. D., Mahony, P. and Offord, R. E., unpublished results).

There is also a number of possible side-reactions (see p. 39), but these are not sufficient to detract seriously from the usefulness of the group.

The acetimidyl group is removed by a strongly ammoniacal reagent (see Example (7.5)). Rather surprisingly, the reagent does not seem to harm a number of proteins.

Maleylation. Maleic anhydride will acylate the $-NH_2$ group (Butler *et al.*, 1969).

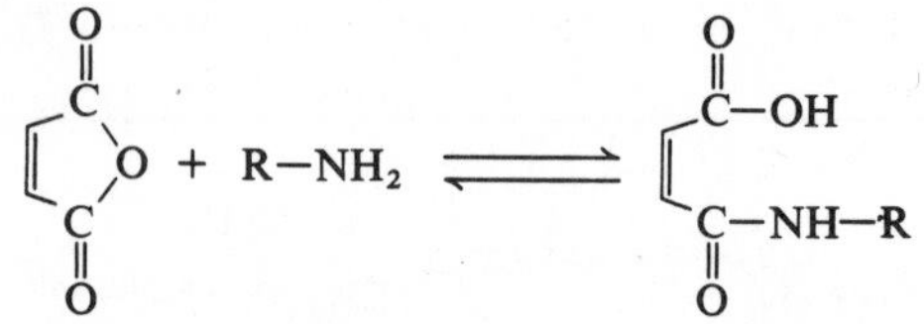

$$\text{cis-}CH(CO)_2O + R{-}NH_2 \rightleftharpoons \text{cis-}HO{-}CO{-}CH{=}CH{-}CO{-}NH{-}R \qquad (2.9)$$

Example (2.6) gives experimental details. The $-SH$ group would also be acylated and it is better to protect it in advance (see Example (2.11)). The *N*-maleyl group is removed in aqueous solution or suspension at pH 3.5 with a half-time at 37 °C of about 11 h (Butler *et al.*, 1969). Rees and Offord (1976a) report a half-time of 14 h. This remarkable acid lability is due to intramolecular catalysis.

Other analogues of maleic anhydride (Table 2.1) have been employed. All are, to a greater or lesser extent, more readily cleaved from the $-NH_2$ group

Table 2.1. Maleic anhydride and some of its analogues.

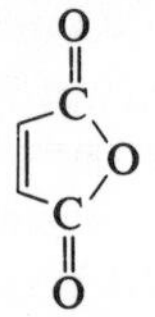

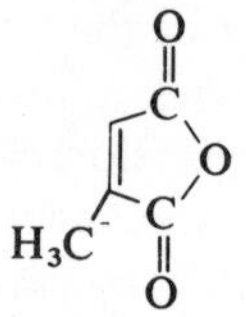

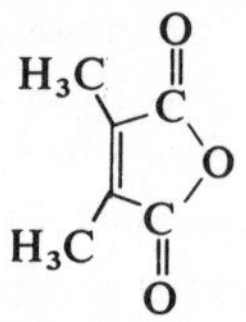

Compound	Reference
maleic anhydride	(Butler *et al.*, 1969)
2-methylmaleic anhydride	(Dixon and Perham, 1968)
2,3-dimethylmaleic anhydride	(Dixon and Perham, 1968)
tetrafluorosuccinic anhydride	(Braunitzer *et al.*, 1968)
exo-cis-3,6-endoxo-Δ^4-tetrahydrophthalic anhydride	(Riley and Perham, 1970)

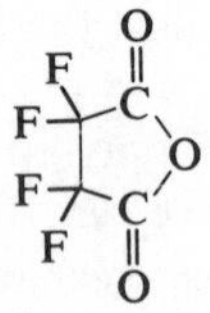

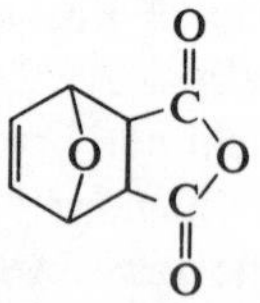

than the maleyl, as a result of enhanced intramolecular catalysis. All but the tetrafluorosuccinic anhydride have been tried in semisynthesis, but Rees and Offord (1976a) report that they have disadvantages in comparison to the unsubstituted maleic anhydride.

Whichever anhydride is used the result is the replacement of the —NH_2 with another functional group, —COOH, which must be protected in its turn. Rees and Offord (1976a, b) report certain instances in which this major disadvantage did not rule out the use of maleylation, but in general the future application of such anhydrides in semisynthesis is likely to be limited.

Selective reactions of —NH_2 groups

Stepwise cleavage from the amino terminus calls for a free α-NH_2 group (Chapter 3). If our reagents protect the α-NH_2 as well as those on side-chains, then —NH_2 protection prior to cleavage is incompatible with stepwise semisynthesis. Although it is generally true that reagents will not distinguish between α- and ε-NH_2 groups, the acetimidyl group does seem to be at least a partial exception. Nureddin and Inagami (1975) showed that, in trypsin, only the side-chain —NH_2 groups are susceptible to acetimidylation. A similar suggestion (though less conclusively verified) had been made about the reaction between the reagent and insulin (Hunter and Ludwig, 1962) and this may be a general phenomenon. Experiments by Garner and Gurd (1975) and by Webster (1974) have provided instances in which the preference for the ε-NH_2 group was only partial, but sufficient to be of some use.

Even where a reagent does not of itself have an absolute preference for the α- or the side-chain groups, specific reaction may occur in a particular molecule due to steric factors.

Example (2.7) gives conditions in which the reaction between Edman's reagent (see Reaction (3.1), p. 49) and the —NH_2 groups of insulin occurs predominantly on the α-NH_2 of the phenylalanine at the amino terminus of the B-chain to give N^{α}-Phe^{B1}-phenylthiocarbamoyl-insulin. After isolation of the B1 derivative from the mixture of products, the other two —NH_2 groups can be protected with acid-stable groups. The phenylthiocarbamoyl group is removed by exposure to strong acid (Reaction (3.2)—and as is well known it conveniently takes with it the residue that it had protected).

By contrast, Example (2.8) gives conditions under which the other two —NH_2 groups of insulin (the α-NH_2 of the glycine at the amino terminus of the A-chain and the ε-NH_2 of lysine B29) are preferentially reactive (Geiger *et al.*, 1971; Geiger and Langner, 1975) (Reaction (2.10) overleaf). The Example shows that several types of protecting group can be selectively applied to insulin in this way.

Even a partial preference may be exploited if the desired derivative can be separated from the products of unwanted acylation. For instance, the precautions described in Examples (2.7) and (2.8) are not absolutely specific, and

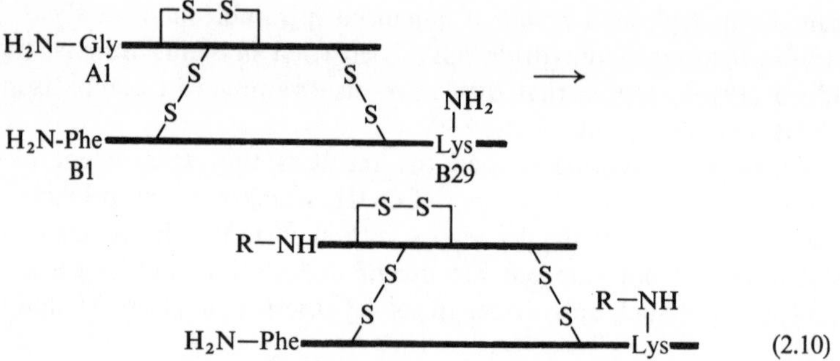

(2.10)

that in Example (2.3) still less so. Example (2.9) gives some methods for resolving the mixtures produced.

Metal-catalysed *transamination* is one of the very few chemical reactions that are obligately specific for the α-NH_2 group. Since it is normally used as part of a method of stepwise cleavage its role in semisynthesis is discussed further in Chapter 3.

Carboxyl protection

Carboxyl protection can also be accomplished before cleavage, but the problem of insolubility of the esterified product usually prohibits our doing so. Nonetheless Example (2.10) describes the acid-catalysed methylation of a protein,

$$R{-}COOH + CH_3OH \underset{}{\overset{[H^+]}{\rightleftharpoons}} R{-}CO{-}O{-}CH_3 + H_2O \qquad (2.11)$$

as it is this group that is at present the most promising for use in prior protection. Offord *et al.* (1976) report preliminary experiments with a reagent that gives an ester which retains the negative charge and which may prove to have analogous advantages to those suggested above for the acetimidyl group (see also Example (4.7)).

Chapter 4 shows that other means besides prior protection are available to solve the problem of distinguishing between α- and side-chain —COOH groups. We shall delay a discussion of the relative merits of the available esterifying groups, including —CH_3, until then.

Protection of cysteine and cystine

The sulphitylation reaction (Swan, 1957) has been the method of S-protection most used in semisynthesis. It is equally applicable to cysteine as to cystine.

$$R_1{-}S{-}S{-}R_2 + 2\,NaHSO_3 \overset{[\text{oxidizing conditions}]}{\rightleftharpoons} R_1{-}S{-}SO_3Na + R_2{-}S{-}SO_3Na + H_2O$$

$$2\,R{-}SH + 2\,NaHSO_3 \rightleftharpoons 2\,R{-}S{-}SO_3Na + H_2O \qquad (2.12)$$

Experimental details are given in Example (2.11).

Although sulphitylation has been advocated for conventional synthesis (Hofmann and Katsoyannis, 1963) it has not been used very much in that way. It is not popular in conventional synthesis because its polar nature reduces the solubility of the product in apolar solvents. This property becomes a definite advantage in semisynthesis where, as has been said, the usual strategy is to keep the protected molecule as hydrophilic as possible.

The —S—SO_3^- group can be removed by reduction or by the action of Hg^{2+}.

$$R{-}S{-}SO_3^- + R_1SH \rightleftharpoons R{-}SH + R_1{-}S{-}SO_3^- \qquad (2.13)$$

The slow acceptance of the group for conventional work may also in part have been due to doubts as to its stability, once applied. However, Rees and Offord (1976a) showed that the group is stable to HBr/trifluoroacetic acid, mild aqueous base, hydrazine, and to organic bases in dimethylformamide.

The —SH groups protected are either natural to the protein in question or have been created by reductive cleavage of disulphide bridges (as we saw a moment ago, the cleavage and sulphitylation are carried out as a concerted process). Stepwise semisynthesis, and sometimes even fragment-condensation synthesis, need not always oblige us to break the —S—S— bridges. They are, for instance, usually kept intact when working on insulin. The absence of a need to reform them by oxidizing mixtures of the chains constitutes a very considerable advantage over conventional synthesis.

Other side-chains: imidazole and indole

We have now dealt with the majority of the operations for side-chain protection that are involved in the minimal protection schemes of semisynthesis. In this Laboratory we protect the imidazole ring of histidine, though only if the molecule is later to be subjected to the action of a diazoalkane (see Chapter 4). One is as likely to carry out such protection either before or after cleavage. The reactivity of the imidazole nitrogen is sufficiently different from that of an —NH_2 group for there to be no problem of differential protection from the α-NH_2 of the fragment. It is possible to protect the imidazole groups of proteins with the Boc—NH—CH(CF_3)-reagent (Reaction (1.21), p. 18). This and other methods of histidine protection will be discussed in the form in which they apply to fragments in Chapter 4.

So many of the operations of peptide synthesis and semisynthesis involve strong acid that it is tempting to seek to protect the indole ring of tryptophan—a notoriously acid-labile group. The indole ring is also sensitive to destruction by the derivatives of protecting groups that are released on cleavage. Rees and Offord (1976a) have examined the formylation reaction of Previero *et al.* (1967) and found that, while it can be applied easily enough, it did not in their hands protect completely against degradation under acid deprotection conditions. Formylation did not, in fact, appear to offer any advantages over careful use of trifluoroacetic acid, as described in the examples in Chapter 7.

EXAMPLES

Example (2.1). The preparation of N^{α}-Boc-Gly^{A1},Phe^{B1},N^{ϵ}-Boc-Lys^{B29} insulin

Source. Offord *et al.* (1976) based on numerous standard methods by other authors.

Process involved. Reaction (1.12a), in which R = $(CH_3)_3C$—.

Method

Insulin (30 mg, 15 μmol —NH_2) was dissolved in redist. dimethylsulphoxide (3 ml) contained in a round-bottomed flask. Redist. triethylamine was added until the externally indicated pH (see p. 217) was between 8 and 9. t-Butyloxycarbonyl azide (0.9 ml, 6.35 mmol) was then added, with stirring to avoid precipitation of the insulin. If necessary, the externally indicated pH was restored to between 8 and 9 with more triethylamine. The mixture was left at room temperature overnight, and then freeze dried. The product was checked by electrophoresis in urea or isoelectric focussing (Examples (A.7) and (A.8)). It travelled in both systems as a substantially homogenous triacyl insulin.

Notes

(1) *N*-Ethylmorpholine can be substituted for triethylamine, particularly if very small quantities of material are involved, if an excursion to too high a pH cannot be risked, or if the pH shows any tendency to wander. It is less easy to overshoot the desired pH with *N*-ethylmorpholine: any modest excess of base simply acts as a reserve of buffering power. See Note 3 to Example (6.6).

(2) If traces of dimethylsulphoxide in the product are objectionable, dimethylformamide can be substituted. It is less easy to dissolve the insulin in the latter solvent unless the protein has previously been freed of Zn^{2+} by dialysis against 1 mM HCl.

Zn^{2+}-insulin can be dissolved in dimethylformamide by the method of Borrás and Offord (1970a, b), which is generally applicable to proteins that are thermodynamically soluble but kinetically reluctant (see p. 218). The method leaves a trace of water in the solvent and can only be used when this can be tolerated.

(3) The excess azide condenses in the cold trap of the freeze drier. Care should be taken when defrosting not to inhale its vapour. It is advisable to run the liquid from the trap into a solution of, say, ethanolamine, in order to destroy the azide.

(4) Benzyloxycarbonyl thiosulphate can be used in a similar way for the protection of —NH_2 groups (Caldwell *et al.*, 1966). This reagent, which is freely

soluble in water, is better for use on proteins than the chloride because of the danger of *O*-acylation by the latter, although this danger can be minimized by the appropriate choice of base (Baer and Eckstein, 1962). H. T. Storey (unpublished work) found that the product of the benzyloxycarbonylation of insulin was much less soluble in aqueous solvents and more difficult to characterize than the Boc- derivative.

(5) Manufacturers have begun to withdraw t-butyloxycarbonyl azide from sale because of fears that it may be exceptionally hazardous. The reagent di-t-butyloxydicarbonate makes an excellent substitute (Moroder *et al.*, 1976). It is more reactive than the azide and can probably be used in the above preparation in a smaller excess. Examples of the use of the dicarbonate in place of the azide are given in Note 5 to Example (2.8) and in Note 3 to Example (4.1).

Example (2.2). The preparation of N^{ϵ}-trifluoracetyl cytochrome *c*

Source. Adapted, with only minor alterations, from the method of Fanger and Harbury (1965).

Process involved. Reaction (1.15).

Method

Cytochrome *c* (100 mg, 150 μmol —NH_2) was dissolved in 4 ml water and brought with vigorous stirring to pH 10 with 1 M KOH. The addition of alkali was done either by hand or by means of an autotitrator. With the reaction vessel in the fume cupboard, ethyl thioltrifluoroacetate (1 ml) was added and the pH, which tended to fall, was kept at 10 by the addition, in small increments with continuous stirring, of further quantities of 1 M KOH. After 50 min, 0.4 ml more of ester was introduced. After a further 40 min, the pH was adjusted to 8.0 with 1 M HCl, and the solution centrifuged. The trifluoroacetyl cytochrome *c* was then precipitated from solution by the careful addition of 1 M HCl (a pH of 3 was sufficient). The suspension was centrifuged and the pellet washed three times with water and dried. The product was checked by tryptic digestion of the CNBr fragments or by end-group analysis (see Examples (A.2) and (A.3)). The reaction appeared to be substantially complete.

Notes

(1) The ethane thiol liberated in the reaction has an unpleasant odour and is toxic. Borrás and Offord (1970a, b), working with insulin, used phenyl trifluoroacetate rather than the thio ester in case the ethane thiol should attack the disulphide bridges. In fact it seems not to do so, at least in anhydrous dimethylformamide solution (Levy and Paselk, 1973).

(2) Fanger and Harbury (1965) dialysed the pH 8 solution (see above) against 0.01 M phosphate buffer, then against water and freeze dried. They then subjected the product to gel filtration. A column of Sephadex G-75 (2.5 cm × 40 cm) in 0.1 M NH_4HCO_3 is suitable. A brown impurity was removed from the red product by this means. The gel-filtration step is useful but may not be essential at this point, since gel filtration is involved at an early stage of any subsequent semisynthetic scheme (see Chapter 3). A trial preparation should help decide.

(3) Several grades of cytochrome *c* are commercially available. We find that grades supplied by reputable manufacturers are suitable even if trichloroacetic acid was used in their preparation, provided that they are quoted as 95% pure or better (based on content of Fe). The principal fault exhibited by such preparations is that they are partially deamidated, but it is rare for significantly more than 10% of the protein molecules to show the loss of even one amide group (D. E. Harris, private communication).

Example (2.3). The preparation of *N*-phthaloyl insulins

Source. Borrás (1972).

Process involved. Reaction (1.17).

Method

Insulin (60 mg, 30 μmol —NH_2) was suspended in 30 ml of a 1% (w/v) aqueous solution of NaCl. The pH was adjusted to 9.5 with 0.05 M NaOH in order to dissolve the protein and then returned to pH 8.5 with 0.1 M HCl. Solid *N*-carboethoxyphthalimide (19.7 mg, 90 μmol) was added and allowed to react for 8 h at 40 °C with stirring and control of the pH to 8.5 by 0.2 M NaOH in the pH-stat. The protein was then recovered from the reaction mixture by adjustment to pH 5.2 with 0.1 M HCl. The pellet was centrifuged, washed three times with water, and dried *in vacuo*. DEAE-cellulose chromatography (see Example (2.9)) was then used to separate the trisubstituted product from partly unreacted material.

Notes

(1) The preparation of the reagent and its use in conventional synthesis is described by Nefkens *et al.* (1960).

(2) The rather small excess of reagent used above gives more of the partially acylated insulins than of the triacyl derivative. These partially protected materials are, however, quite useful. Geiger and Langner (1975) describe conditions under which the principal product is the N^{α},N^{ε}-phthaloyl-Gly^{A1},Lys^{B29} derivative (see Note 3 to Example (2.8)).

Example (2.4). The preparation of N^{α}-Gly^{A1},N^{α}-Phe^{B1},N^{ϵ}-Lys^{B29}-tri(N^{α}-Cbz-methionyl) insulin

Source. Adapted from Levy and Carpenter (1967).

Process involved. Reaction (2.5), in which R = $C_6H_5CH_2$—O—CO—.

Method

Zn^{2+}-free insulin (60 mg, 30 μmol —NH_2) was dissolved in redist. dimethylformamide (5 ml) and sufficient triethylamine was added to bring the externally indicated pH (p. 217) to between 9 and 10. Cbz-methionine hydroxysuccinimido ester (125 mg, 300 μmol; see Example (5.5)) was added, and the externally indicated pH adjusted if necessary. The reaction was allowed to proceed for 18 h at 20 °C. The product was then precipitated from solution with ether, collected by centrifugation, washed twice with ether, and dried *in vacuo*. The product was checked by end-group analysis (see Example (A.2)). No free —NH_2 groups were detected.

Notes

(1) Levy and Carpenter (1967), when coupling Boc-methionine *p*-nitrophenyl ester, used insulin hydrochloride, added 10 mg triethylamine, and did not find it necessary to check the pH.

(2) A urethan derivative of methionine, as used in this Example, is preferable to a simple acyl derivative (see Note 3 to Example (7.7)). Ethyloxycarbonylmethionine is a better group if there is to be a long exposure to acid (see Note 2 to Example (3.2)).

(3) The S atom in the side-chain of methionine is liable to oxidation. The resulting derivative will not undergo the deprotection reaction with CNBr (Reaction (2.6), Example (7.7)). It does not seem to be necessary to conduct all the operations under N_2, but long-term storage of the derivative should take place in sealed tubes under N_2 at −20 °C.

Example (2.5). Acetimidylation of the —NH_2 groups of chicken triosephosphate isomerase

Source. Milman (1974), based on the method of Hunter and Ludwig (1962).

Process involved. Reaction (2.7). The reaction for the preparation of the reagent (Note 1) is

$$CH_3CN + CH_3OH + HCl \rightleftharpoons CH_3C\begin{matrix} \nearrow\!\!\!\!= NH.HCl \\ \searrow OCH_3 \end{matrix}$$

Method

Triosephosphate isomerase (100 mg, 112 μmol —NH_2) was dissolved in 50 ml of 4 M urea. The urea solution was freshly deionized (p. 217). Methyl acetimidate hydrochloride (0.35 g, 3.2 mmol) was added in six separate, roughly equal portions at intervals of 20 min. The pH fell as the reagent was introduced and was adjusted to pH 9.5 by the manual addition of 5 M NaOH (very small drops, with vigorous stirring). Once the pH had reached 9.5, it was kept at this value by the addition of 2 M HCl from the autotitrator until the time arrived for the next addition of the reagent.

The mixture was dialysed in the cold room in Visking 18/32 tubing against three changes of distilled water over the course of 2 days. The product was checked through the estimation of N^{ε}-acetimidyl lysine by amino-acid analysis (Note 4, Example (A.1)). The reaction had gone substantially to completion.

Notes

(1) Although methyl and ethyl acetimidate hydrochloride are commercially available, it is more economical to prepare the reagent onself. The following preparation of the reagent is that of Hunter and Ludwig (1962), with additional precautions due to Milman (1974) and D. E. Harris (unpublished work). Hydrogen chloride (25 g dry gas) was dissolved in methanol (20 ml) at −40 °C or below. Acetonitrile (20 ml) was added quickly to the cooled solution. After 30 min the mixture was warmed up to 0 °C in an ice bath.

The reaction starts while the mixture is warming up, or sometimes after it has stood for a while at 0 °C. The reaction liberates heat, and there can be a sudden and dangerous increase of pressure in the flask. It is therefore important that (a) the contents of the flask should occupy considerably less than half of its volume; (b) the flask should be agitated in an ice bath during the warming-up period, so as to control any sudden rise in temperature; and (c) the warming-up should be carried out in a fume hood, and the flask subsequently left there for several hours (or even overnight). Many workers prefer to make provision for the relief of any pressure that builds up in the flask during these operations; this must be done via some system that prevents the entry of moisture.

When the flask had stood for some hours at 0 °C, it was stoppered, and kept at 4 °C for 3 days. The resulting solid was ground with a pestle and mortar under methanol/dry ether (1: 2, v/v), filtered, and washed with ether. The material was sealed for storage after removal of the ether *in vacuo*. In our hands the methyl acetimidate has m.p. 86–88 °C (decomp.), although Hunter and Ludwig reported 93–95 °C (decomp.), but is perfectly satisfactory for protein modification. Bates *et al.* (1975) found that the reagent is stable for at least a year when stored dry at −20 °C.

(2) An alternative method of pH control involves the pre-adjustment of a solution of the reagent to pH 9.5 with 5 M NaOH.

(3) With a protein of this size, the Visking tubing probably need not be boiled (cf. Note 2 to Example (2.6)).

(4) The difference between the pK of the acetimidyl group and that of the —NH_2 group is generally such that it is possible to remove any partially substituted protein by ion-exchange chromatography (e.g. Di Marchi *et al.*, 1978). These authors also draw attention to the existence in the reaction of molecules with one or more lysine side-chains blocked by a group that suppresses the charge.

Example (2.6). The preparation of N^{α}-, N^{ϵ}-maleyl,*S*-sulphityl lysozyme

Source. Rees and Offord (1976a) based on Butler *et al.* (1969).

Process involved. Reaction (2.9), with prior protection of —SH groups by sulphitylation (Example (2.11)).

Method

Sulphityl lysozyme (Example (2.11)) was dissolved in freshly deionized 8 M urea (p. 217) to a concentration of 10 mg/ml (5 μmol —NH_2/ml). The urea solution was 0.2 M in sodium tetraborate buffer, pH 8.5. The mixture was cooled to 4 °C and stirred magnetically. (The temperature was kept below 4 °C throughout the subsequent reaction.) Small portions of powdered maleic anhydride (recryst. from chloroform) were added. The pH fell at each addition and was restored to pH 8.5 (apparent value given by a glass electrode, standardized at 20 °C against a urea-free buffer) by 2 M NaOH. The operation was repeated until a 12-fold molar excess of reagent over —NH_2 groups had been added. The solution was stirred, still at 4 °C, for 1 h after the apparent pH had been returned to 8.5 following the last addition of anhydride. The pH remained constant during this time. The solution was dialysed in pre-boiled Visking 18/32 tubing (Note 2) against three changes of 20 times its volume of NH_4HCO_3 solution (0.5%, w/v) at 4–10 °C. The solution was then freeze dried. The product was checked by tryptic digestion and by reaction with 1-fluoro-2,4-dinitrobenzene (Example (A.3)). The reaction appeared to have gone substantially to completion.

Notes

(1) Rees and Offord (1976a) report that the use of other, related anhydrides, higher excesses of maleic anhydride, or higher temperatures can lead to undesirable side-reactions. If the reaction is carried out as described, it does not seem to be necessary to treat with NH_2OH (Riley and Perham, 1970) to remove maleyl groups from side-chain —OH groups.

(2) Visking 18/32 dialysis tubing has the smallest pore size of the types that are easily available. Nonetheless, some losses of small proteins occur, particularly when they have been modified in such a way as to prevent aggregation

or to give an open, flexible conformation—for example mono- and higher acyl insulins escape, while native insulin does not (Borrás, 1972). In the present instance *N*-maleyl,*S*-sulphityl lysozyme escapes, while the native and *S*-sulphityl enzymes do not. The pore size can be reduced by exposure to boiling $NaHCO_3$ (1%, w/v) solution (a technique in use at the National Institutes of Health, Bethesda). To control the process exactly it is necessary to plunge the tubing into a beaker of the solution which is already boiling and, after the chosen interval, remove it to a beaker of cold water. The tubing is kept immersed in the boiling liquid by pressure from a plastic vessel of appropriate diameter, laid on the surface of the solution. Contact between the tubing and sharp-edged objects should be avoided. The pore size is progressively reduced with time of exposure and 10 min is found to be sufficient to prevent the escape of any of the modified proteins described in this Chapter (but see Note 3 to Example (3.5)). Treated tubing is not kept from day to day.

This procedure is normally used in this Laboratory even when the protein is large, since the material extracted from the tubing during boiling imparts a yellow colour, and an unpleasant smell, to the solution. It seems prudent to remove it as a matter of routine.

(3) If traces of NH_4HCO_3 in the final product are objectionable it is necessary either to omit it from the dialysis medium or to freeze dry for at least 48 h with an operating pressure of less than about 0.1 Torr (13 Pa). There seems to be little benefit in redissolving the dried solid and refreezing.

Example (2.7). The preparation of N^{α}-Phe^{B1}-phenylthiocarbamoyl insulin

Source. A combination of the methods of Brandenburg (1969), Borrás and Offord (1970a, b), Borrás (1972), and Saunders (1974).

Process involved. Reaction (3.1), p. 49. The reaction is largely, though not exclusively, directed to the α-NH_2 of phenylalanine B1. Glycine A1 and lysine B29 may be less reactive because of steric factors and the ε-NH_2 of lysine B29 has, in addition, a less favourable p*K* for reaction under the conditions described. Burton and Lande (1970, 1971), and Lande (1971) have discussed the additional selectivity likely to be observed when —NH_2 groups have different degrees of nucleophilicity. Unreacted material and other unwanted products are nonetheless found in the reaction mixture and must be separated from the main component (Example (2.9)).

Method

Insulin (760 mg, 380 μmol —NH_2) was suspended in distilled water (21.1 ml) at 4 °C and cold pyridine (42.2 ml) added slowly with stirring and cooling. The protein dissolved during the addition. Redist. phenyl isothiocyanate (27.4 μl, 230 μmol) was dissolved in pyridine (21 ml) and this mixture was added, after cooling, to the insulin solution. After 16 h at 4 °C the pyridine was removed in a rotary evaporator under reduced pressure. The volume of the solution was kept roughly constant by the intermittent addition of water. When the majority of the pyridine had been removed, the solution was adjusted to pH 5.1 with 6 M HCl. The precipitated protein was recovered by centrifugation and washed once with water before being dried *in vacuo*. The reaction product was examined by isoelectric focussing (Example (A.8)) and found to contain 15–20% unreacted insulin, about 60% monophenylthiocarbamoyl insulin, and 20–25% diphenylthiocarbamoyl insulin. The mono-derivative fraction was approximately 85% of the N^{α}-Phe^{B1} isomer. Very little tri-derivative was detected. The reaction products were separated by ion-exchange chromatography (Example (2.9) Method A).

Notes

(1) Borrás and Offord (1970) described the control of the reaction in a dilute aqueous medium by the autotitrator. The balance of products was very similar. Borrás (1972) gave curves of the extent of formation of the different products as a function of time for both methods of control (Fig. A1.4, p. 208).

(2) Brandenburg (1969) gave spectral data for insulin, the mono-derivative and the di-derivative. Borrás (1972) reported that insulin and the mono-, di-, and triphenylthiocarbamoyl derivatives have characteristically different ultraviolet absorption spectra, and that the concentration of a sample of a purified derivative is given by the relationship

$$\text{concentration (mg/ml)} = K \times (\text{optical density at 275 nm})$$

$K = 1.00$ for insulin; 0.41 for mono-derivatives; 0.23 for di-derivatives; and 0.12 for the tri-derivative.

(3) Brandenburg (1969) separated the products by gel chromatography and, in approximately 50 mg lots, by gel electrophoresis.

(4) The S atom in the phenylthiocarbamoyl group is liable to oxidation. This must be avoided, as the oxidized group is unable to participate in Reaction (3.2), p. 49. It is therefore best not to store the derivative for any length of time. Some workers prefer to carry out the reaction after vigorous purging of the solutions and reaction vessel with N_2, particularly if it is proposed to stir the reaction mixture (cf. Note 4 to Example (3.2)). The product has also been stored under N_2 at -20 °C.

(5) Care must be taken to avoid bumping during the rotary evaporation, particularly towards the end of the the operation, when precipitated protein is present.

Example (2.8). The preparation of N^{α}-Gly^{A1},N^{ϵ}-Lys^{B29}-di-Boc- insulin, and analogous derivatives with other protecting groups

Source. Adapted by D. J. Saunders from Geiger *et al.* (1971), Geiger and Langner (1971) and R. Geiger (private communication).

Process involved. Reaction (2.10) (R = $(CH_3)_3C—O—CO—$) limited to two of the three $—NH_2$ groups of insulin by kinetic factors. See Notes 3 and 4 for details of the other derivatives.

Method

Insulin (250 mg, 125 μmol $—NH_2$) was suspended in 10 ml fresh aqueous $NaHCO_3$ (5%, w/v) and stirred at room temperature. Methanol (40 ml) was added slowly, with cooling if necessary. The insulin had dissolved by the time that the addition of the methanol was half completed. A slight precipitate of $NaHCO_3$ occurred toward the end of the addition but was ignored. t-Butyloxycarbonyl azide (0.625 ml, 4.35 mmol) was added and the reaction was allowed to continue for 16 h at 20 °C. A quantity of 6 M HCl, equivalent to the $NaHCO_3$, was then added carefully, and the bulk of the methanol removed in the rotary evaporator under reduced pressure. Care was taken to prevent frothing. The sample was not allowed to dry out completely: 20 ml of water was added halfway through this process. The pH was then adjusted to 4.8 with 1 M HCl. The protein was collected by centrifugation, washed twice with water, three times with acetone, once with ether, and dried in air. The product was examined by isoelectrophoresis (Example (A.8)). The balance between the various protected products is discussed in Note 2, below, and their separation from each other in Example (2.9).

Notes

(1) Geiger *et al.* (1971) used dimethylformamide as the organic solvent. Geiger (private communication) has indicated that the best method is to use methanol as the organic component, and to control the pH by the addition of 2 M NaOH in methanol from an autotitrator. The method of isolation of the product given above differs in minor respects from that of Geiger *et al.*

(2) Geiger *et al.* (1971) reported that the N^{α}-Gly^{A1},N^{ϵ}-Lys^{B29} di-derivative accounted for 85% of the product. The remainder was N^{α}-Gly^{A1} mono-derivative (6%) and tri-derivative (9%). Saunders and Offord (1977b) give conditions under which approximately 50% of the starting weight of insulin is converted to N^{α}-Boc-Gly^{A1} insulin. It was also possible to isolate a small amount of the N^{ϵ}-Lys^{B29}mono-derivative. The main difference from the method given above was that the solvent was aqueous triethylamine (5%, v/v)/methanol, 1:4 (v/v), adjusted to an apparent pH (glass electrode, 20 °C) of 9.3. The reaction took place for 3 h at 20 °C. The two predominating mono-derivatives can be separated from each other by means of the ion-exchange system described in Example (2.9), Methods A and B.

(3) Geiger and Langner (1975) reported that under the above conditions similar selectivity is shown in the reaction between insulin and *N*-carboethoxyphthalimide (cf. Example (2.3)).

(4) Saunders and Offord (1977a) described the selective acylation of the $—NH_2$ groups of residues A1 and B29, using the above conditions but with

ethyloxycarbonylmethionine hydroxysuccinimido ester (Note 4, Example (5.5)) used in place of the azide. A ratio of 2.5 : 1 over —NH_2 groups gave the greatest proportion of the di-derivative, in which the N^{α}-Gly^{A1},N^{ϵ}-Lys^{B29} form predominated. Using Cbz-methionine hydroxysuccinimido ester (which behaves in a similar manner to the ethyloxycarbonyl derivative when used at 2.5 : 1 over —NH_2 groups), they found that a ratio of 1 : 3 gives roughly 50% unreacted insulin and 50% mixed mono-derivatives, while a ratio of 8 : 1 gives 25% tri-derivative and 75% di-derivatives.

(5) As mentioned previously, the azide can be replaced by smaller quantities of di-t-butyloxydicarbonate. Moore (1978a) used a 9-fold excess of the reagent over —NH_2 groups (27 μl for 250 mg insulin) in the preparation of the di-derivative. He used a quantity equimolar to $-NH_2$ groups for the preparation (Note 2, above) that maximizes the yield of mono-derivatives. He used the solvent mixture described in Note 2 for both types of preparation. The dicarbonate seems less satisfactorily selective than the azide when using des-Phe^{B1}-insulin as the starting material (cf. Example (3.1)).

(6) Geiger *et al.* (1975a) used similar conditions for selective acylation with methylsulphonylethyloxycarbonyl-*N*-hydroxysuccinimido ester. In a bulk preparation (which can be very considerably scaled down, A. R. Rees and M. Whittle, private communication) they dissolved 5 g of insulin (approximately 2.5 mmol —NH_2) in a mixture of 10 ml of 1 M $NaHCO_3$ and 20 ml dimethylformamide. Some $NaHCO_3$ was precipitated. A further 40 ml dimethylformamide was then added. Then, at 45 min intervals, were added 200, 112, 100 and 50 mg of the hydroxysuccinimido ester (a total of 462 mg, 175 mmol). After a further 45 min stirring the pH was adjusted to approximately 5 with acetic acid. The crude product was precipitated by ether as an oil. After decanting off the ether, it was taken up in a little methanol and 5 vol of ethyl acetate added. The insoluble material was filtered, washed with ethyl acetate and ether, and dried *in vacuo* over P_2O_5 and KOH. The subsequent purification of the product was by gel filtration on Sephadex G-10 (recovery of protein 75%), and partition chromatography as described in Example (2.9), Method C. 480 mg of the di-derivative were obtained in this way from 1 g of crude product, with much smaller quantities of mono- and tri-derivative. The di-derivative was shown to be virtually pure N^{α}-Gly^{A1},N^{ϵ}-Lys^{B29} substituted.

The removal of the group is discussed in Example (7.3).

Example (2.9). The separation of the products of selective protection of the —NH_2 groups of insulin

Source. Methods A and B: Borrás (1972), Saunders (1974) based on the work of numerous authors especially Bromer and Chance (1967) and Brandenburg *et al.* (1972). Method C is adapted from Geiger and Langner (1973).

Process involved. Methods A and B: ion-exchange chromatography. Method C: partition column chromatography.

Method A

Microgranular DEAE-cellulose (Whatman DE-52, 50 g wet wt.) was equilibrated according to the manufacturers' instructions with 0.01 M Tris-HCl/0.03 M NaCl/7 M urea, adjusted to pH 7.7 (glass electrode, 20 °C) by the addition of 6 M HCl. The exchanger was packed into a column (2.8 cm × 38 cm) placed in the cold room and pumped with buffer at 40 ml/h until the bed volume remained constant. The reaction product from Example (2.7) was dissolved in 15 ml of buffer and the resulting drop in pH counteracted by the addition of the required quantity of 1 M NaOH. The sample was pumped into the column bed and a linear gradient started between 1 litre of starting buffer and 1 litre of similar buffer in which the concentration of NaCl was 0.07 M. The gradient was pumped through the column at 40 ml/h. The components emerged at approximately the following molarities of NaCl: unreacted insulin, 0.045 M; N^{α}-Phe^{B1} mono-derivative, 0.052 M; N-Gly^{A1} mono-derivative, 0.054 M; di-derivative (predominantly N^{α}-Gly^{A1},N^{α}-Phe^{B1}), 0.061 M. The peak tubes were pooled, adjusted to pH 4.8 and dialysed at 4 °C in boiled Visking tubing, 18/32 (see Note 2 to Example (2.6)) against large volumes of distilled water. The protein derivatives precipitated during dialysis and were collected by centrifugation and dried *in vacuo*. Spectroscopy showed that very little was left in the supernatants.

Method B

All operations were carried out at 20 °C. DEAE-Sephadex, A-25, was equilibrated with buffer (0.1 M Tris/0.05 M NaCl/7 M urea adjusted with 6 M HCl to pH 8.4 (glass electrode). It was packed into a column to form a bed (2.4 cm × 13 cm) and pumped at 60 ml/h until the bed volume remained constant. The same flow rate was used during the separation. The product from Example (2.8) (150 mg) was dissolved in 15 ml of buffer that had not been adjusted with HCl. The solution was then adjusted to pH 8.4 by very careful addition of 6 M HCl and pumped into the column. A linear gradient was then started between 400 ml of the pH 8.4 buffer as described above and 400 ml of an otherwise similar buffer in which the concentration of NaCl was 0.25 M. The components of the mixture emerged at approximately the following concentrations of NaCl: unreacted insulin, 0.065 M; N^{α}-Gly^{A1} mono-derivative, 0.076 M; N^{ϵ}-Lys^{B29} mono-derivative, 0.11 M;N^{α}-Gly^{A1},N^{ϵ}-Lys^{B29} di-derivative, 0.133 M; tri-derivative, 0.15–0.16 M. Protein was obtained from the peak tubes as described in Method A.

Method C

Water (1.8 l), butan-1-ol (360 ml), and glacial acetic acid were shaken vigorously together for 15 min and then allowed to stand overnight. Sephadex LH-20 (30 g) was stirred manually for 10 min with 30 ml of the upper, organic phase of the solvent mixture. The resulting damp powder was then subjected to gentle mechanical stirring with 160 ml of the lower, aqueous phase. The slurry was poured into a column (38 cm × 1.8 cm; bed volume 97 ml) and packed under gentle air pressure provided by a balloon, or large syringe. Equilibration was completed by pumping lower phase at 40 ml/h for 2 days. The bed should be kept continuously packed by flow under suction as it tends to float in the solvent. The Cbz-methionyl product from Example (2.8) was dissolved in lower phase (120 mg in 5 ml), drawn into the column and eluted with more of the lower phase. Insulin emerged after about 50 min, mono-derivatives after about 70 min, di-derivatives after about 2 h, and tri-derivative after about 3.5 h. (Tri-derivatives dissolve poorly in the eluant.) The peaks showed some cross-contamination by their neighbours. This problem was worst with the tri-derivative and scarcely noticeable with the di-derivative. Once the derivatives had emerged, the column was ready for re-use. The derivatives were recovered from the peak tubes by lyophilization. The overall recovery from the column was better than 90%.

Notes

(1) The choice between Methods A and B is best made on the basis of experiment. Method C works best for derivatives of very apolar protecting groups. (For instance, Method C is not very successful with the ethyloxycarbonyl-methionyl product, although the resolution might improve sufficiently if a longer column was used (cf. Geiger and Langner, 1973).)

(2) The column in Method A has a greater axial ratio than is usually considered necessary in true ion-exchange chromatography. A more compact column (4 cm diam. × 6 cm height) gave good resolution of the insulin, mono- and di-derivative peaks from each other, but did not resolve the two isomeric mono-derivatives from each other very well (Saunders, 1974).

(3) The urea solutions used in Methods A and B were made by diluting the freshly deionized solution described on p. 217. Should the buffer still contain small quantities of finely particulate material, the particles will accumulate on the top of the column. If the passage of buffer is continued for too long before the application of the sample, there may be sufficient accumulated deposit to bind to the sample to a noticeable extent and interfere with the elution. If there is no visible discoloration at the top of the column, this problem will be unlikely to be serious. But it should be kept in mind as a possible explanation if unforeseen difficulties arise.

(4) The purpose of the adjustment of pH of the pooled fractions in Methods A and B is to increase the efficiency of precipitation during dialysis. If this is done, losses on dialysis are negligible and the overall recovery of protein from the column is about 75–80%.

(5) Because of their previous treatment, the samples for separation are acidic to a greater or lesser degree—hence the special care taken to control the pH of the sample solutions in Methods A and B.

(6) The Sephadex LH-20 must be stirred with care, in order to avoid damage to the structure of the gel heads.

(7) The eluant in Method C tends to separate into two phases on even a small drop in temperature.

(8) Larger samples can be separated in Method C but the eluate tends to gel, or even solidify, as the peak emerges from the column. Geiger and Langner (1973) used a column 4 cm × 1–2 m for quantities between 0.5 g and 2 g and one 2.5 cm × 100 cm for 300–500 mg.

(9) There is no increase in resolution on slowing the rate of elution in Method C to 10 ml/h. The resolution is not impaired at a flow-rate of 80 ml/h, but the back-pressure then tends to give rise to leaks.

Example (2.10). The preparation of N^{ϵ}-acetimidyl cytochrome *c* methyl ester

Source. Adapted from Chibnall *et al.* (1958) by Wallace (1976).

Process involved. Reaction (2.11).

Method

Dry HCl was bubbled through anhydrous methanol with cooling for about 5 min, with the aim of producing a solution about 0.9 M in HCl (checked by titration with 0.2 M NaOH, using bromothymol blue). If the actual strength was significantly different, further passage of HCl, or dilution with the required volume of anhydrous methanol, was used as appropriate to bring it near to that value. 250 mg (approximately 275 μmol —COOH) of N^{ϵ}-acetimidyl cytochrome *c* (prepared by the method given in Example (2.5)) were dissolved in approximately 10 ml of the fresh solution. The precise volume was chosen so that, when more anhydrous methanol was added to give a final volume of 100 ml, the solution would be 0.1 M in HCl. Some precipitation of protein occurred, but all went back into solution during reaction, with stirring, for 18 h at 20 °C. The solution was then dried by rotary evaporation. The product was checked by gel electrophoresis and by examination of the tryptic digest. (See Example (A.4): a successfully protected sample will not give any migrating peptides.) The reaction appeared to have gone substantially to completion.

Notes

(1) The methanol was made perfectly dry by the Grignard method (e.g. Vogel, 1967) and stored over molecular sieve Moisture was excluded from the reaction mixture.

(2) With some substrates (but not cytochrome *c*) the rotary evaporation step may well help the reaction go to completion.

(3) An alternative method (Wallace, 1976) is to suspend the protein (or peptide, see Chapter 4) in 15 ml methanol. The methanol should be of Analytical Reagent grade, but need not be specially dried. There should then be added

0.125 ml of 12 M HCl (Analytical grade, or better). The resulting solution should be stored at 20 °C for 2 h and then dried by rotary evaporation. If the dampness of the solution can be tolerated, this method is more convenient than that given above.

Example (2.11). Simultaneous reduction and sulphitylation of the disulphide bridges of lysozyme

Source. Rees and Offord (1976a).

Process involved. Reaction (2.12).

Method

Hen-egg lysozyme was dissolved in freshly deionized 8 M urea (p. 217) to a concentration of 2 mg/ml (0.57 μmol disulphide bridges/ml). A solution of 1 M Tris-HCl, pH 7.5, was added until the Tris concentration was 0.1 M. Anhydrous Na_2SO_3 was dissolved in the solution to a final concentration of 0.3 M. Oxygen was bubbled through the solution for 20 min and, if necessary, the pH adjusted to 7.5 with 1 M HCl. Cysteine hydrochloride was dissolved in the mixture to a final concentration of 1.6×10^{-4} M. The solution was then transferred to a microbiological incubator rack and shaken, loosely stoppered, at 37 °C for 18 h.

At the end of the reaction the solution was dialysed in the cold room in Visking 18/32 tubing against three changes of 20 times its volume of distilled water over the course of 2 days. The non-diffusible residue was then freeze dried. The product was checked by examining a tryptic digest (see Examples (A.4) and (A.5), and Rees and Offord, 1976a). All the cysteine-containing peptides were found to have the sulphityl group.

Notes

(1) The most satisfactory commercial grades of lysozyme tend to be those that are readily soluble in distilled water and free of salt.

(2) The conditions given above are more vigorous than some in the literature (e.g. Swan, 1957; Chan, 1968). Rees and Offord (1976a) found that lysozyme resisted these milder versions of the technique.

(3) It is not essential to pre-boil the dialysis tubing (Note 2 to Example (2.6)). The protein precipitates during the dialysis.

(4) Some sequence workers believe that proteins that have been extensively handled are cleaved less efficiently by CNBr. If true, the reason is likely to be the partial S-oxidation of methionine residues (cf. Note 3 to Example (2.4)). Milman (1974) considered it prudent to carry out CNBr cleavage before sulphitylation because of the possibility of oxidation during the latter reaction.

Whether this precaution was necessary or not, the procedure gave quite satisfactory cleavage and protection in that instance. However, the converse danger —that CNBr could oxidize free —SH groups—should also not be overlooked. The choice of strategy should be made on the basis of trial experiments, with thorough characterization of the products.

CHAPTER 3

Cleavage and separation of fragments

Cleavage is the characteristic operation of protein semisynthesis. It exemplifies very well the fact that semisynthesis depends, both conceptually and practically, on the methodology of amino-acid sequence determination as well as on that of conventional peptide synthesis.

Sequence determination, too, naturally has its stepwise and its fragment approaches. In either of these the agent that cleaves the peptide bond may be chemical or enzymic.

Stepwise cleavage: the Edman reaction

Chemical agents are by far the most popular for stepwise cleavage: the Edman reaction (Edman, 1950) is pre-eminent. Residues are removed from the amino terminus in two stages. First the —NH_2 group reacts with phenyl isothiocyanate.

$$\mathrm{C_6H_5{-}N{=}C{=}S + H_2N{-}\underset{}{\overset{R_1}{\overset{|}{C}}H}{-}CO{-}NH{-}\overset{R_2}{\overset{|}{C}}H{-}CO{-}NH\ldots \longrightarrow}$$

$$\mathrm{C_6H_5{-}NH{-}\underset{\underset{S}{\|}}{C}{-}NH{-}\overset{R_1}{\overset{|}{C}}H{-}CO{-}NH{-}\overset{R_2}{\overset{|}{C}}H{-}CO{-}NH\ldots} \qquad (3.1)$$

Then the derivative is treated with strong acid—trifluoroacetic acid is as popular in sequence determination as it is in synthesis—and the amino-terminal residue is removed by cyclization.

$$\mathrm{C_6H_5{-}NH{-}\underset{\underset{S}{\|}}{C}{-}NH{-}\overset{R_1}{\overset{|}{C}}H{-}CO{-}NH{-}\overset{R_2}{\overset{|}{C}}H{-}CO{-}NH\ldots \longrightarrow}$$

$$\mathrm{C_6H_5{-}NH{-}C{=}N{-}CH(R_1){-}C({=}O){-}S\ (ring) \; + \; H_2N{-}\overset{R_2}{\overset{|}{C}}H{-}CO{-}NH\ldots} \qquad (3.2)$$

If the peptide possesses ε-NH_2 groups, they are able to participate in Reaction (3.1) but Reaction (3.2) is not possible. In the time that Reaction (3.2) takes to remove the amino-terminal residue, about 10% of the ε-NH_2 groups are regenerated. The rest remain as the ε-phenylthiocarbamyl derivatives.

As is well known, the advantage of the Edman reaction is that Reaction (3.2) liberates a new α-NH_2 group which can react in its turn according to Reaction (3.1). Each pair of Reactions (3.1) and (3.2) constitutes a cycle of the process (cf. stepwise synthesis, p. 19).

The actual determination of sequence can be done by the identification of the phenylthiohydantoin; by discovering what has been lost from the amino-acid composition of the peptide after each cyclization step; or by using any of the standard methods of identifying the new amino-terminal residue on a sample taken after each cycle. As with stepwise synthesis, the process lends itself to automated repetition (Edman and Begg, 1967). Also, as in stepwise synthesis, the stepwise degradation is not the whole answer for very long sequences because of the diminishing returns expressed by the $(q/100)^{n-1} \times 100$ relationship (p. 20) and the increasing heterogeneity due to the random incidence of failed cycles.

The Edman reaction has adapted well to semisynthesis. Much the same considerations apply as do when it is used for sequence work, although the quantities of protein or peptide involved in a cleavage for a semisynthesis are likely to be considerably larger. If only one or a few cycles of the process are to be carried out, it is possible to rectify the consequences of any small departures from 100% efficiency of the various steps by chromatographic purification at some intermediate stage. It is necessary to protect ε-NH_2 groups before cleavage, both to avoid N^{ϵ}-phenylthiocarbamylation and to permit the truncated fragment to be used subsequently in a controlled coupling. We have seen that, in general, protection of the ε-NH_2 groups of a protein cannot be effected without protecting the α-NH_2 as well. Since Reaction (3.1) calls for a free α-NH_2 group, a protein can only be prepared for stepwise semisynthesis by the Edman reaction when steric factors permit us to distinguish chemically between α- and ε-NH_2 groups (as was true in Examples (2.7) and (2.8)) or when we can take advantage of the partial preference of methyl acetimidate for the ε-NH_2 group (see p. 31 and Garner and Gurd, 1975). No such problems arise if we wish to use the Edman reaction on fragments prepared by endolytic agents after prior protection of the —NH_2 groups of the protein. All such fragments, except the one that arises from the amino terminus of the protein, have free α-NH_2 groups and are suitable for the Edman reaction without further treatment. We therefore have a powerful means of modifying a chosen fragment in fragment-condensation semisynthesis and several Examples are given in Chapter 5. Examples (3.1) and (3.2) give details of the use of the Edman reaction to remove a few residues from a protein.

Gurd and coworkers (e.g. Di Marchi *et al.*, 1978a) use the 3-sulphophenyl isothiocyanate. The resulting derivative of the peptide or protein has an additional negative charge by virtue of the sulpho group. This property can be exploited in an ion-exchange separation of the product of Reaction (3.2) from any material that might fail to cyclize.

Stepwise cleavage: transamination

Valuable though the Edman reaction is, its use entails treating the protein or peptide with a strong acid, often with a considerable cumulative exposure. Many molecules seem to suffer no harm, but to have an alternative, more gentle method would be an advantage. Dixon (see Dixon and Fields, 1972) has called attention to transamination as a possible means of removing residues under mild, aqueous conditions. Each cycle has two stages: transamination of the α-NH_2 group to an α-oxo group

glyoxylate + protein + M^{2+} metal ⇌ … ⇌ glycine + α-oxo-protein

(3.3)

and then scission.*

(3.4)

* The Scheme is only an outline one. For the many alternative pathways see Dixon and Fields (1972).

We can protect the ε-NH_2 groups between Reactions (3.3) and (3.4), thereby solving the problems of differential protection which arise when seeking to apply the Edman reaction to proteins. There has been one report of the use of transamination for a semisynthetic purpose (Webster and Offord, 1972) and this forms the basis of Example (3.3). Mild and straightforward though the reaction may appear, puzzling failures of step (3.4) have occurred (D. Webster and R. E. Offord, unpublished results; C. Herrera-Caravajal, unpublished results). Some of the products of the process have proved hard to identify—though admittedly in a difficult system (Webster, 1974). The method and others like it, such as the cobalt enamine method (Collman and Buckingham, 1963), have great promise and hold out the main hope of controlled stepwise cleavage of proteins that are very sensitive to acid, but more development is certainly required.

Stepwise cleavage: exopeptidases

Enzymes can in principle be used for the stepwise shortening of chains. Leucine aminopeptidase (which operates on the amino terminus) and the carboxypeptidases (which operate on the carboxyl terminus) are likely to be among the most useful, although they do not remove all residues with equal ease. The principal problem in their use is the fact that, if we interrupt the digestion after a given time, we are likely to find that some molecules have been degraded further than others and we have a heterogeneous product. The further the digestion proceeds the more serious this danger becomes. However, the enzymes sometimes remove a number of residues quantitatively and then come to a stop, leaving the next residue untouched. The molecules that happened to lag behind in their degradation then catch up and we once again have a homogeneous product.

If we are to use the products of exopeptidase digestion for subsequent coupling reactions, we are still confronted with the same problems of differential protection between α- and side-chain that we encountered when discussing the use of the Edman reaction.

We shall see in Chapter 8 that the carboxypeptidases have been used, in a most elegant way, to produce semisynthetic proteins. Carboxypeptidase A and carboxypeptidase B (which has the convenient property of restricting its operations to the removal of lysine and arginine residues) have been used on fragments that had been given differential protection of the side-chain —COOH groups before digestion commenced (see Chapter 5).

Endolytic agents and fragment-condensation semisynthesis

So far the products of stepwise cleavage have been used, naturally enough, mainly in the stepwise form of semisynthesis. A few residues are removed from one end and replaced by stepwise coupling of aminoacyl active esters. Fragment-condensation semisynthesis has made use of the products of endolytic agents, some of which we must now consider.

Trypsin

Trypsin is to be preferred among the proteolytic enzymes because of its tendency to promote total cleavage adjacent to just two types of residue (lysine and arginine). Most other proteases neither restrict themselves to such a small number of residues nor do they always cleave 100% at the sites that they choose. The products of digestion with such less specific enzymes are much more complex mixtures of fragments than if trypsin is used, though this is not to say that other proteases of reasonably sharp specificity could not be used with success.

Trypsin promotes the cleavage of peptide bonds formed by the α-COOH groups of lysine and arginine, which together account on average for about 10% of the residues in proteins of known sequence (Dayhoff, 1972). The modal length of tryptic peptides is therefore about 10 residues, though with an observed range from 1 residue to over 100.

Limited tryptic cleavage

The number of points of cleavage can be reduced, and the length of the fragments thereby increased, by chemical modification of one of the two types of side-chain. Very few side-chain derivatives of arginine or lysine will make satisfactory contacts in the specificity site of trypsin and so its ability to promote cleavage is suppressed. In practice it is usually the ε-NH_2 group of lysine that is modified and cleavage is then restricted to arginine. Proteins of known sequence contain on average about 4% arginine (Dayhoff, 1972) and so the modal length of the fragment is about 25 residues.

All the —NH_2-protecting agents mentioned in Chapter 2 also suppress tryptic cleavage at lysine residues—a powerful additional reason for our carrying out —NH_2 protection before cleavage.

The sequence of the protein may itself impose a further degree of limitation on the cleavage, as trypsin will not cleave -Lys-Pro- or -Arg-Pro- bonds. It is, also, reluctant to cleave bonds of the type (Lys(or Arg)-Asp(or Glu) and the existence of such a sequence in cytochrome *c* enabled Harris and Offord (1977) to obtain larger fragments for semisynthesis than would otherwise have been possible.

Example (3.4) gives conditions for tryptic cleavage which are generally applicable and Example (3.5) indicates the general approach when one wishes to carry out proteolytic cleavage of a rather heavily protected, and therefore often quite insoluble, protein. Mixtures of water and organic solvents must then be used, preferably with Ca^{2+} to stabilize the trypsin. Trypsin ceases to be able to catalyse the cleavage of peptide bonds when the percentage of water in the solvent falls to somewhere between 70% and 50%. In Chapter 4 we shall make use of its ability to continue to function as an esterase in solvents where water contents are as low as 30%.

Cyanogen bromide

Cyanogen bromide cleaves the peptide bond at methionyl residues only (Gross and Witkop, 1961).

$$
\begin{array}{c}
\mathrm{H_2N{-}{-}{-}NH{-}CH(CH_2{-}CH_2{-}S{-}CH_3){-}CO{-}{-}{-}COOH} \\
\big\downarrow \ \mathrm{CNBr} \\
\mathrm{H_2N{-}{-}{-}NH{-}\overset{}{CH}{-}C(=O)} \ (\text{ring: } \mathrm{CH{-}H_2C{-}CH_2{-}O{-}C}) \ + \ \mathrm{H_2N{-}{-}{-}COOH}
\end{array}
\qquad (3.5)
$$

Although it is normally most convenient to use strong aqueous acid as a solvent for the reaction, with all the dangers of hydrolysis that that involves, the method is very popular in sequence work. It deserves to be equally popular in semisynthesis as methionine is quite scarce in most proteins. The fragments will be long, which means fewer coupling steps will be needed to re-form the protein. The average abundance is 1.6% (Dayhoff, 1972) and the modal length of the fragments is therefore about 60 residues, although once again there is a large spread of actual values.

Example (3.6) gives details of some typical digestions.

We shall see in later Chapters how it is possible to exploit the bonus presented by the release of the carboxyl-terminal residue of the fragments as homoserine lactone. The formation of the lactone constitutes both a ready-made activation of the α-COOH for coupling and the distinction between α- and side-chain —COOH groups which we so constantly require. We shall see in Chapter 5 how we can attempt to decide in advance whether or not the substitution of homoserine for methionine can be tolerated in our resynthesized protein, and how, if it is not, methionine can be restored in its place.

Now that we have established which are the best agents for cleavage, we can calculate approximately how many residues in an average protein it will be easy to modify semisynthetically. The figures given above for the abundance of amino acids in a typical protein allow us to calculate that in a molecule of, say, 130 residues there will be, on average, five sites for limited tryptic cleavage and two for CNBr cleavage. If, as a working assumption, we say that it would be fairly easy to use the Edman reaction to take up to four residues from any of the new amino termini liberated, we have $(5+2)\times 4 = 28$ residues, any one of which could be substituted with reasonable ease. We could gain easy access to four more, if we assume that we would be able to use the Edman reaction at the original amino terminus of the protein (p. 49). If we could cleanly remove one residue on average from each of the new carboxyl termini liberated, plus one from the original carboxyl terminus of the protein, a further eight residues might be fairly easily accessible. Therefore, out of a total of 130 residues, we can be reasonably confident of being able to modify about 40, that is, one residue in three. This is less satisfactory than being able to reach any of them, as conventional synthesis promises (in theory) to allow, but we can take some comfort in comparing the number of couplings necessary in the two types of approach. We would need to do 129 in the conventional synthesis. The required number for the semisynthesis varies between one (if the site to be modified is at the amino terminus of the protein) and nine (if the site to be modified is four residues within a tryptic fragment), Fig. 3.1 illustrates the point. If we are prepared to discard a fragment, and substitute one for it that is totally synthetic, the number of sites accessible to semisynthesis increases, but of course so does the number of couplings.

Atherton and Sheppard (1978) point out that the most difficult part of conventional synthesis is the coupling of large protected fragments, together with the subsequent deprotection of the product. They suggest that semisynthesis by fragment condensation described above meets the same problems as in conventional synthesis but 'under the less favourable circumstances dictated by the inflexibility of the fragment-preparation procedures'. It is certainly true that one is not so free to choose the point of cleavage and subsequent recoupling, and that one should always strive to reduce the number of couplings, but the sites available for cleavage will not necessarily always be the more disadvantageous ones. Nor, in many cases, will such disadvantages as there might be outweigh the advantages of having the fragments ready-made, and rapidly and lightly protected. Atherton and Sheppard also point out, rightly, that it is unrealistic to expect that side-chain protection reactions will proceed with absolute specificity when applied to complex polyfunctional proteins and their fragments. On the other hand, it might be that the side-reactions would be insignificant, especially in comparison to those that occur during the much more extensive chemical manipulation of conventional synthesis. The best guide to the balance of advantages and disadvantages between the two approaches will be given by comparing the numbers of well-characterized, useful analogues of proteins that each produce, and the quantities in which they are obtained.

Separation of fragments

Most conventional methods of column chromatography will work well for fragments that are minimally protected by groups that are not too apolar. Example (3.7) describes the fractionation of some typical CNBr digests by means of gel filtration in aqueous media. No Example is given of the use of ion-exchange chromatography to separate mixtures of fragments, although it is a most powerful approach. The details vary too much from problem to problem. One can consult the manufacturers' instructions relating to particular exchangers, or reviews such as that by Morris and Morris (1976) in order to learn

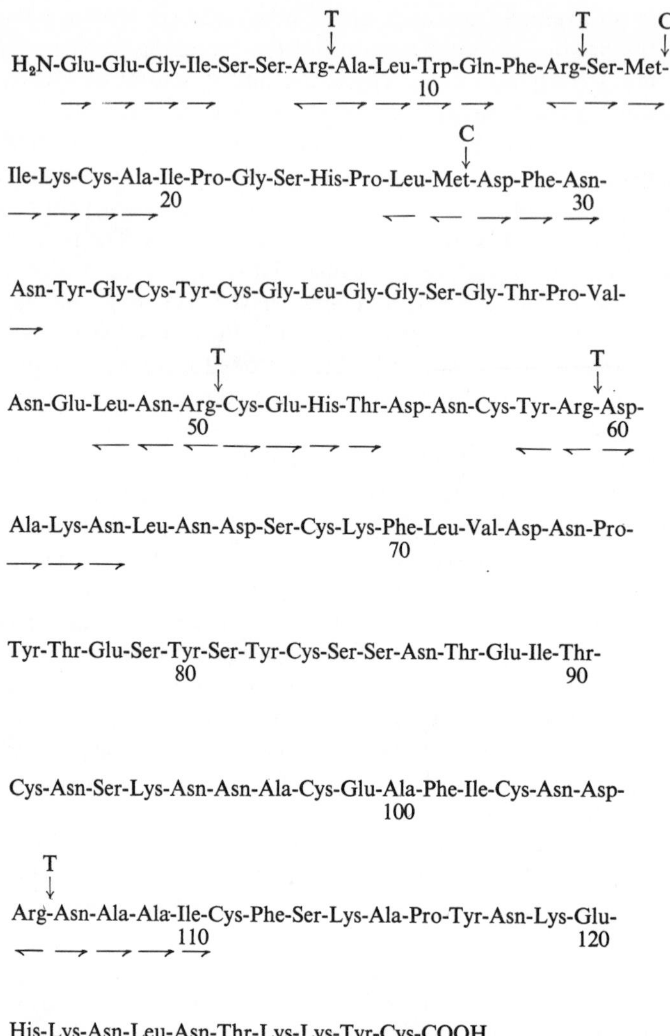

Fig. 3.1. The accessibility of the residues in a typical protein to semisynthetic modification. T = sites for tryptic cleavage after N^{ε}-protection; C = sites for CNBr cleavage; ⇀ = possible removal of residues from fragments by up to four cycles of the Edman reaction; ↽ = possible removal of residues by carboxypeptidases (assuming the usual sort of specificity shown by these enzymes). It will be seen that 40 (30%) of the residues can be removed, and are thus susceptible to modification. The minimum number of coupling steps would be one, the maximum would be 9. It might well be easier to produce some of the analogues of peptides 1–7 and 8–13 by total synthesis, rather than by semisynthetic manipulation, in which case the number of accessible residues rises to 43. If the same principle were applied to changes in fragments 16–27 and 51–59 the number becomes 52 (40%). The molecule has the average arginine and methionine content of those globular proteins that have been sequenced (Dayhoff, 1972).

the general principles of the design of such separations, and then (with a considerable residual element of trial and error) fit them to the particular problem in hand. The set of systems described by Rees and Offord (1976a) is fairly typical of what is required when trying to separate about a dozen fragments for semisynthetic use.

Method A in Example (3.7) is, in fact, based on one developed to separate the CNBr fragments in the sequence determination of an enzyme. Provided, as is true of the acetimidyl group, that the protection does not greatly alter the characteristics of the peptides, it is often well worthwhile looking back for guidance on one's separation methods to the experimental sections of the papers describing the determination of the amino-acid sequence of the protein in question.

When a protein is extensively protected, aqueous media will not usually be of much use. Example (3.8) describes the separation of a tryptic digest of a protected derivative of insulin by gel filtration in organic solvents. Success in this instance is probably due to there being so few components in the digest. Not all such digests can be separated at present and it is to be hoped that other means of separating materials in organic solvents will be brought into use in semisynthesis.

Quite useful quantities of small peptides can be isolated from digests by preparative paper electrophoresis (e.g. Offord, 1969a).

EXAMPLES

Example (3.1). The Edman degradation of N^{α}-Gly^{A1},N^{ϵ}-Lys^{B29}-di-Boc- insulin

Source. A minor adaptation by Saunders (1974) of the method of Geiger and Langner (1973).

Process involved. Reactions (3.1) and (3.2). Reaction (1.13) also occurs under the conditions of Reaction (3.2).

Method

The product of Example (2.8) (60 mg, 10 μmol —NH_2) was dissolved in 3–5 ml pyridine/water (3 : 1, v/v). Redist. phenyl isothiocyanate (9.8 μl, 120 μmol) was added and the mixture allowed to stand for 12–24 h at 4 °C. The protein was removed as described in Example (2.7) except that the pH was taken to 4.5. The dried material was then dissolved in anhydrous trifluoroacetic acid

(10 mg/ml) and allowed to stand for 1 h. After partial removal of the trifluoroacetic acid on the rotary evaporator, the protein was precipitated from solution with ether, washed with ether, and dried *in vacuo*. The product of both stages could be checked by electrophoresis (Examples (A.7) or (A.8)) and end-group determination (Example (A.2)). Both stages normally went substantially to completion. Some idea of the integrity of the three-dimensional structure of the final product could be obtained by a determination of biological activity or by crystallization (Example (7.10); Brandenburg, 1969).

Notes

(1) Scavengers were not used with the trifluoroacetic acid, but it was re-distilled before use.

(2) The ether precipitate must be washed with care. The ether will dry off very quickly, leaving the protein to dissolve to a high, and possibly damaging, concentration in the small amount of trifluoroacetic acid that remains. For the first wash in particular, the ether must be added as soon as the supernatant is removed.

(3) Geiger and Langner (1973) repeated four cycles of the degradation, re-protecting the —NH_2 groups of Gly^{A1} and Lys^{B29} by the method of Example (2.8) between each cycle. They found that the preference of the protecting reagent for residues A1 and B29 persisted as the B-chain was shortened. Nonetheless, the di-protected derivative had to be purified from under- and over-reacted products (Example (2.9)) after each protection. Geiger and Langner do not report, after the first two cycles, whether or not any of the other isomeric di-derivatives begin to appear.

As mentioned in Note 5 to Example (2.8), the di-t-butyloxydicarbonate can replace the azide when applying the Boc- protection. In our hands (R. E. Offord and C. G. Bradshaw, unpublished work) the carbonate seems much less selective than the azide when the des-Phe^{B1}-insulin is protected, in contrast to the satisfactory result obtained with insulin itself.

Example (3.2). The stepwise truncation of the B-chain of N^{α}-Gly^{A1},N^{ϵ}-Lys^{B29}-di-protected insulin by repeated cycles of the Edman reaction

Source. Saunders and Offord (1977a).

Process involved. Reactions (3.1) and (3.2).

Method

N^{α}-Gly^{A1},N^{ϵ}-Lys^{B29}-di-Cbz-methionyl insulin (Note 4 to Example (2.8)) was dissolved in 7 M urea (p. 217) to a concentration of 1–2 mg/ml. Redist. phenyl isothiocyanate (1 μl per mg of protein, a molar excess of 50 over —NH_2 groups) was then added and the reaction allowed to proceed under N_2 for 7–16 h at 20 °C with control of the apparent pH (glass electrode) at 8.75 by the addition of 0.2 M NaOH from an autotitrator. Vigorous mechanical stirring was used to emulsify the reagent, which is insoluble in water. Some protein precipitated during the reaction, and at the end the pH was adjusted to 4.6 with 1 M HCl to bring down more. Approximately 75% of the protein was precipitated, and was centrifuged off, washed with two changes each of water, acetone, and ether before drying *in vacuo*.

The remainder of the protein could be precipitated from the supernatant by dialysis against distilled water. It was recovered and washed in the same way as the original precipitate. The two fractions were combined.

The protein was then dissolved in anhydrous trifluoroacetic acid (10 mg/ml for 1 h at 20 °C), precipitated by ether, washed with ether, and dried *in vacuo*.

The cycle of operations was repeated four times to produce the des-(B1–B4) insulin derivative.

Notes

(1) See Notes 1 and 2 to the previous Example.

(2) Saunders and Offord (1977a) and Saunders (1974) report that the Cbz-groups in N^{α}-Gly^{A1},N^{ϵ}-Lys^{B29}-di-Cbz-methionyl insulin are insufficiently stable to repeated treatment to trifluoroacetic acid. There is a small, but noticeable deprotection at each cycle. The ethyloxycarbonylmethionyl group (Note 4 to Example (2.8)) does not suffer from this disadvantage. Also, the ethyloxy-carbonylmethionyl analogue of the derivative used in the method above reacts more easily with the Edman reagent. 130 mg of the derivative (22 μmol —NH_2 groups) was dissolved in pyridine/water (19 : 1, v/v) to a concentration of 17 mg/ml. Phenyl isothiocyante (20 μl, 168 μmol) was added and the reaction allowed to proceed at 20 °C for 7–16 h. The reaction mixture was then worked up as described in Example (3.1) and used for repeated cycles of degradation.

Geiger and Langner (1975) have used the phthaloyl group with success for the preparation of des-(B1–B4) insulin, but found that the method of deprotection (Example (7.6)) destroyed the des-(B1–B5) derivative. The methylsulphonyl-ethoxycarbonyl group (Note 6 to Example (2.8)) has also been used for these purposes, and works very well (see p. 167).

(3) A larger excess of phenyl isothiocyanate can be used without visible harm if, for example, the protein derivative is valuable and one wishes to guard at all costs against under-reaction.

(4) The vigorous stirring appears to promote the oxidative attack on the phenylthiocarbamoyl derivative (Note 4 to Example (2.7)) and the reactions were carried out under N_2.

(5) Single and repeated Edman degradations are the most common means of cleavage used in published semisyntheses (see Chapter 8).

(6) Residue B4 is glutamine and tends to cyclize to pyrrolidone carboxylic acid once it is exposed at the amino terminus. This damaging side-reaction was

noticeable in the reaction sequence of Geiger and Langner (1973) mentioned in Note 3 to Example (3.1), probably because of the handling necessitated by the protection and separation stage after the third Edman cycle. See also Note 3 to Example (5.2).

Example (3.3). The transamination and scission of the amino-terminal isoleucine residue of bovine trypsin

Source. Webster and Offord (1972) and Webster (1974), based on the methods described by Dixon and Fields (1972).

Process involved. Reactions (3.3) and (3.4). The side-chain $—NH_2$ groups are acetimidylated after the removal of the α-NH_2 group in Reaction (3.3). The new $—NH_2$ group revealed by Reaction (3.4) is thus the only unprotected one in the molecule and is a site for further specific reaction.

Method

Trypsin (10 mg) was dissolved in 10 ml 1 M pyridine/50 mM acetic acid/10 mM nickel(II) diacetate/0.2 M sodium glyoxylate/10 mM $CaCl_2$ and incubated at 37 °C. The protein began to precipitate almost at once and was completely precipitated after 4 h. The suspension was dialysed. If the dialysis was prolonged for 4–5 days the protein went back into solution. The extent of transamination was checked by the 2,4-dinitrophenylhydrazine method of Fields and Dixon (1971) and by end-group analysis (Example (A.2)). The reaction appeared to have been complete after 4 h. The material was then acetimidylated by the method of Example (2.5).

The scission step was carried out by incubating in 0.1 M acetic acid/20 mM *o*-phenylenediamine/10 mM $CaCl_2$ for 18 h at 37 °C. The scission, as judged by the 2,4-dinitrophenylhydrazine test and end-group analysis, was about 90% complete. The material was then dialysed.

Notes

(1) The sodium glyoxylate was prepared from commercial glyoxylic acid by the method of Radin and Metzler (1955). *o*-Phenylenediamine was five times recryst. from ethyl acetate.

(2) The $CaCl_2$ in the transamination and scission mixtures is present solely to stabilize the trypsin and would in general not be needed. In spite of its presence, the enzyme loses activity in parallel with the extent of transamination, whether at 37 °C or (with slower reaction) at 4 °C. Activity is not restored on scission. Controls suggest that the enzyme is not seriously damaged by the individual components of the reaction mixture. This conclusion is supported by experiments involving the transamination, scission, and activation of trypsinogen.

(3) The transaminated protein was brought back into solution simply by dialysis for a long period of time. If this was not done, solid urea had to be added to the solution for the 2,4-dinitrophenylhydrazine test. Similarly, the end-group analysis had to be carried out in the presence of 8 M urea (Note 2 to Example (A.2)).

(4) Isoleucine residues are resistant to transamination and require the strong reaction conditions given above. Other residues can be reacted under milder conditions (see Dixon and Fields, 1972 and, for example, Sarkar *et al.*, 1978).

(5) The acetimidylation reaction (Example (2.5)) has at least a partial specificity for ε-NH_2 groups. It could therefore have been employed before transamination, although it seemed better not to do so.

(6) In a control experiment the more usual scission reagent (2 M sodium acetate/2 M acetic acid/40 mM *o*-phenylenediamine) damaged samples of the native enzyme. The milder solution that was actually used seemed to do no damage and to be effective in promoting the scission.

(7) As stated in the text, this reaction probably needs further development before it can be used with confidence. The scission step, in particular, seemed to lead to side-reactions when applied to insulin (C. Herrera, unpublished work). If the scission step were really effective, then active esters of α-oxo acids would make excellent —NH_2-protecting reagents (Saunders, 1974).

Example (3.4). The tryptic cleavage of *N*-maleyl,*S*-sulphityl lysozyme

Source. A standard method. The particular details are taken from Rees and Offord (1976a).

Process involved. The specific catalysis by trypsin of the hydrolysis of peptide bonds formed by the —COOH groups of arginine. The lysine residues are blocked against the action of the enzyme by N^{ϵ}-maleylation.

Method

N-Maleyl,*S*-sulphityl lysozyme (Example (2.6)) was dissolved in NH_4HCO_3 solution (0.5%, w/v) to a final concentration of 10 mg/ml. The solution was warmed to 37 °C. Solid trypsin was dissolved in the solution: one-hundredth of the weight of lysozyme was used. The digest was stopped after 90 min at 37 °C by freeze drying. The extent of digestion was checked by the separation of the peptides (Example (A.4)). Little or no undigested material remained.

Notes

(1) The protein derivative, possibly because of the extended conformation imposed on it by the many negatively charged substituents, is extremely susceptible to cleavage by the chymotryptic activity present in most samples of

trypsin. The combined effect of the two activities is, in particular, to liberate a peptide Val-Ala-Trp-Arg (corresponding to residues 109–112 of the sequence). This peptide migrates on paper electrophoresis at pH 6.5 (Example (A.3)) with a mobility of +0.38 (Asp = −1) and stains for tryptophan. Its presence in a digest of the lysozyme is an extremely sensitive and specific test for the chymotryptic activity. Such non-tryptic cleavage makes it necessary to prepare, or obtain from a reputable supplier, trypsin treated with a chymotrypsin inhibitor such as L-1-chloro-4-phenyl-3-(*p*-toluenesulphonylamide)-2-butanol. Even with this precaution, it is advisable to restrict the digestion to as short a period as possible.

(2) The shortness of the digest makes it unnecessary to stabilize the trypsin by the addition of Ca^{2+} (cf. Example (3.5)).

(3) The digestion could be followed in an autotitrator, rather than being controlled by the buffer (cf. Example (4.8)).

Example (3.5). The tryptic cleavageof $N^{\alpha,\epsilon}$-tri-Boc-,N^{im}-di-Boc—NH—CH(CF_3)-insulin

Source. H. T. Storey and R. E. Offord, unpublished work.

Process involved. As for Example (3.4). There is only one arginine residue, that at position B22 (Ryle *et al.*, 1955).

Method

The insulin derivative (prepared as described in Examples (2.1) and (4.10)) was dissolved in redist. dimethylformamide to a concentration of 5 mg/ml. To this solution was added an equal vol. of *N*-ethylmorpholine (2%, v/v)/0.04 M $CaCl_2$, adjusted to pH 8.6 with conc. HCl. Trypsin (one-tenth of the weight of the insulin) was dissolved in 0.02 M HCl to a concentration of 20 mg/ml and added to the insulin solution. The mixture was left for 18 h at 20 °C and then dialysed at 4 °C against several changes of distilled water (2 litres each) for 8 h. The sample was then freeze dried. The extent of digestion was checked by paper electrophoresis at pH 6.5 (see Example (A.4)) and after gel filtration (see Example (3.8)), by amino-acid analysis of the separated peaks.

Notes

(1) The high concentration of dimethylformamide is necessary to dissolve the insulin derivative. High concentrations of other organic solvents, such as dimethylsulphoxide and pyridine, have also been used in tryptic digests of highly protected molecules. Trypsin does not work well as a protease in such solvent mixtures even with Ca^{2+} in the buffer—indeed the concentration of water in the present instance is near to the lowest practical limit—and a larger than usual

enzyme–substrate ratio is required. The esterolytic activity of trypsin is able to tolerate a decrease in the concentration of water somewhat better (see Example (4.8)).

(2) The trypsin must, as in Example (3.4), be freed of chymotryptic activity.

(3) The dialysis tubing, Visking 18/32, was boiled as described in Note 2 to Example (2.6). The duration of treatment was extended to 30 min, in order to close the pores of the tubing to the point at which the octapeptide Gly-Phe-Phe-Tyr-Thr-Pro-Lys(N^{ϵ}-Boc)-Ala (corresponding to residues B23–B31) is retained in good yield, along with the des-(B23–B31)-N^{α}-di-Boc,N^{im}-di-Boc—NH—CH(CF_3)-insulin.

(4) It would no doubt be possible to remove the trypsin by passage through a column of an appropriate trypsin inhibitor bound to a solid phase. The eluant would have to be sufficiently aqueous not to suppress the binding of enzyme to inhibitor..

Example (3.6). Cleavage of N^{ϵ}-acetimidyl cytochrome *c* and of N^{ϵ}-acetimidyl cytochrome *c* methyl ester by CNBr

Source. Wallace (1976).

Process involved. Reaction (3.5). See Note (4) to Example (3.7) for a description of the nature of the fragments obtained.

Method

The derivative of cytochrome *c* (prepared as described in Examples (2.5) or (2.10)) was dissolved in 0.1 M HCl at a concentration of 10 mg/ml. Solid CNBr was added in a 60-fold molar excess over methionyl residues, see Note 5. The mixture was allowed to react for 20 h at 20 °C and then freeze dried. The extent of cleavage was assessed by separation of the components of the digest on gel filtration (see Example 3.7)).

Notes

(1) Esterification by the methyl group leaves the behaviour of the protein unaltered both in the cleavage and in the subsequent separation of the fragments.

(2) Cyanogen bromide has an unpleasant, toxic vapour. Unwanted CNBr is best destroyed by 1 M NaOH; the resulting solution contains NaCN. Cyanogen bromide should be handled in the fume hood and weighed in a stout, sealed bottle. The contents of the cold trap of the freeze drier should be run into 1 M NaOH when defrosting after a CNBr digest has been freeze dried.

(3) The CNBr is usually obtained in the form of fairly large, clear crystals. These can be broken up by introducing a dry magnetic follower into the weighing bottle, and causing the follower to tumble (rather than spin) by placing the

bottle on a magnetic stirrer run at a higher than normal speed. Old samples of CNBr often change to an orange powder and there may be a build-up of pressure in the stock bottle. We discard bottles containing such material, although a test suggested that it is just as effective at promoting the cleavage as a fresh, crystalline CNBr.

(4) Other acidic solvents can be used for the cleavage. A 70% (v/v) solution of formic acid in water is often employed for cytochrome *c* (e.g. Corradin and Harbury, 1970). An example of its use on another protein, $N^{\alpha,\epsilon}$-acetimidyl triosephosphate isomerase (Milman, 1974), is as follows. Protein (1.29 g, prepared as described in Example (2.5)) was dissolved in 100 ml of 70% formic acid. Solid CNBr (304 mg, a 27.4-fold excess over methionyl residues) was dissolved in the solution and the reaction allowed to proceed at 20 °C. A further 304 mg of CNBr was added after 24 h. Six hours after this, 25 ml water was added and the reaction was terminated by freeze-drying lyophilization (the sample did not stay frozen, but did not froth). The dry residue was resuspended in 25 ml water and redried to remove traces of reagent.

(5) Black and Leaf (1965) treated cytochrome *c* as in the method above, but with a 15-fold excess of CNBr over methionyl residues. The digest then contained more of the longer products that arise from conversion of methionine residues to homoserine without cleavage of the peptide chain. Wallace (1976) studied the formation of these by-products in 70% formic acid with a protein concentration of 15 mg/ml. With a 25-fold excess of CNBr over methionine residues, he found that the proportion of the longer products in the final mixture was much reduced, but that with a 1.5-fold excess, the longer products were obtained in high yield, with their methionine residues intact (cf. Note 4 to Example (3.7)).

Example (3.7). Separation of CNBr fragments of proteins

Source. Method A: Milman (1974), Furth *et al.* (1974), based in part on Corran and Waley (1974). Method B: Wallace and Offord (1979) based on Corradin and Harbury (1970).

Process involved. Method A: Solvent extraction followed by gel filtration on Sephadex G-75. Method B: Gel filtration on Sephadex G-50.

Method A

The digest of $N^{\alpha,\epsilon}$-acetimidyl triosephosphate isomerase (Note 4 to Example (3.6)) was suspended in 25 ml 0.025 mM dithiothreitol and adjusted to pH 8–9 with 1 M NH_4OH. The mixture was stirred at 20 °C for 14 h and the insoluble material removed by centrifugation. The pellet was washed with water and the washings combined with the supernatant and freeze dried (fraction I). The pellet was then stirred for 18 h at 20 °C with 10–20 ml 0.025 mM dithiothreitol/8 M urea. The mixture was then adjusted to pH 3 by the addition of 1 M HCl and dialysed for 28 h against three changes of 4 litres 0.001 M HCl/0.01 mM dithiothreitol. The non-diffusible residue was freeze dried (fraction II).

Fraction I was dissolved in 7.5 ml acetic acid/pyridine/water (300 : 16 : 684 by vol., pH 3.2) and applied to a column of Sephadex G-75 in the same solvent (135 cm × 1.6 cm) and run under a gentle, gravity-fed head of pressure. Three well-resolved peaks were observed at 280 nm, numbered 1, 2, and 3 in order of emergence. The peak tubes were pooled and freeze dried. Fraction II was suspended in 100 ml of the pyridinium acetate solution, the mixture stirred and centrifuged. The pellet was washed with more of the pyridinium acetate solution and the washings combined with the supernatant. Both the supernatant and pellet fractions were then freeze dried. The supernatant fraction was run on the G-75 column as before and a further quantity of peak 1 was obtained.

Method B

The freeze-dried digest of N^{ϵ}-acetimidyl cytochrome *c* or that of the methyl ester was taken (250 mg) and dissolved in 10 ml of 7% (v/v) aqueous formic acid. The solution was applied to a column (135 cm × 3.5 cm) of Sephadex G-50 run in the same solvent, and eluted at a flow rate of 80 ml/h. When the digest from a 120-fold excess of CNBr (Wallace, 1976) was run, seven peptide peaks were observed at 280 nm. Their K_d values were computed. ($K_d = (V_e - V_0)/(V_t - V_0)$, where V_e = the volume at which the peptide is eluted, V_t = the total volume of solvent taken up by the gel, and V_0 = the volume at which peptides elute that are so large as to be totally excluded from the gel beads.) The values were: peak 1, 0.0; peak 2, 0.12; peak 3, 0.19; peak 4, 0.28; peak 5, 0.37; peak 6, 0.56; peak 7, 0.68; peak 8, approximately 1.0. Peaks 1, 2, 3, and 4 had the colour of haem. The peaks were imperfectly resolved, but could be obtained reasonably free of their neighbours after freeze drying and separate reapplication to the column.

Notes

(1) Triosephosphate isomerase has 248 residues, with methionyl residues at positions 13 and 82 (Furth *et al.*, 1974). Amino-acid analysis and end-group determination showed that peak 1 was the 69-residue peptide corresponding to residues 14–82, peak 2 was the 13-residue peptide corresponding to residues 1–13. The final pellet from fraction II was a quite pure sample of the 166-residue fragment corresponding to residues 83–248. Peak 3 did not contain any peptide material.

(2) The treatment with dithiothreitol in Method A was to try to prevent the —SH groups of the two larger peptides from cross-linking by the formation of disulphide bridges. Fraction I was removed before the urea treatment because it contained a peptide that might have been lost during the dialysis needed to remove the urea (but see Note 3 to Example (3.5)).

(3) The smallest fragment (Acim-Ala-Pro-Arg-Lys(Acim)-Phe-Phe-Val-Gly-Gly-Asn-Trp-Lys(Acim)-homoserine) does not require any further side-chain

protection for semisynthesis. The largest fragment, similarly, has no free —NH_2 group except the α-NH_2 of its amino-terminal residue, and is ready to act as an amino component in a coupling—one would not normally wish to see it as a carboxyl component since it is the carboxyl-terminal fragment of the protein. (It is considerably larger than most of the molecules ever synthesized by conventional means.) The middle fragment could be used as an amino component, but requires further treatment if it is to be used as a carboxyl component.

(4) Horse-heart cytochrome *c* has 104 residues and has methionyl residues at positions 65 and 80 (Margoliash *et al.*, 1961). Amino-acid compositions show that peak 2 represents the full 104-residue sequence and peak 1 a polymer of it. Although its mode of formation is not known, the polymer has often been seen in work on cytochrome *c* (Margoliash and Schejter, 1966; Corradin and Harbury, 1970). Peak 3 represents residues 1–80; peak 4, residues 1–65; peak 5, residues 66–104; peak 6, residues 81–104; and peak 7, residues 66–80. Peak 8 did not contain peptide. Peak 6 is ready for use as an amino component, or for stepwise truncation without further protection. It is not necessary to rechromatograph peaks 6 and 7 straightaway, as they are usually well resolved, and in any case the next stage of their treatment normally involves a gel-filtration step. The amino-terminal residue of fragments 66–80 and 66–104 is glutamic acid, and excessive handling will cause it to cyclize (cf. Note 6 to Example (3.2) and Note 3 to Example (5.2)).

Peaks 1, 2, 3, and 5 represent a failure of the reagent to cleave at one or more methionyl residues. The amino-acid analyses of these materials show that internal methionyl residues have been wholly or partly converted to homoseryl residues, but without cleavage. The causes for this well-known phenomenon of CNBr digestion have been discussed (e.g. Carpenter and Shiigi, 1974). The amount of conversion of methionine can be made quite small (Corradin and Harbury, 1970)—cf. Note 5 to Example (3.6).

(5) A plot of K_d vs. log (molecular weight) for the peptides in Method B is linear (Wallace, 1976), as would be expected from the work of Andrews (1965). The possession of a haem group does not appear to affect K_d. The two tryptic peptides of N^ϵ-acetimidyl cytochrome *c* corresponding to residues 1–38 and 39–104, have K_d values in this system of 0.38 and 0.25, respectively (Harris, 1977). These points lie on the same line as the others.

(6) Electrophoresis of the smaller peptides showed that those with carboxyl-terminal homoserine were predominantly in the lactone form when isolated on the formic-acid column. The fragments in Method A were predominantly in the open form, as a result of the extractions at pH 8. Note 1 to Example (4.1), and Example (4.9), indicate the methods by which the lactone and open forms can be interchanged.

(7) The digests used in this Example happen to have only a few fragments of rather different lengths. More complex mixtures would probably require ion-exchange chromatography for their separation.

Example (3.8). Separation of the tryptic digest of $N^{\alpha,\epsilon}$-tri-Boc,N^{im}-di-Boc—NH—CH(CF_3)- insulin

Source. H. T. Storey and R. E. Offord, unpublished work.

Process involved. Gel filtration by lipophilic Sephadex LH-20.

Method

The digest from Example (3.5) (50 mg) was dissolved in 2 ml dimethylformamide and applied to a column of Sephadex LH-20 (1.8 cm × 75 cm). The column was eluted with dimethylformamide/water (17 : 3, v/v) at 20 ml/h. Two peaks were observed at 280 nm and were freeze dried.

Notes

(1) Amino-acid and end-group analysis showed that the first peak to emerge was trypsin and the des-(B23–B31)-N^{α}-di-Boc,N^{im}-di-Boc—NH—CH(CF_3)-insulin and the second was the octapeptide corresponding to residues B23–B31. Tests showed that the non-volatile salts from the digestion emerged after the second peak. The insulin derivative was obtained in a pure form from the first peak by dissolving the freeze-dried material in 1 ml dimethylformamide and adding 10 ml water, with cooling. The precipitate of insulin derivative was collected by centrifugation, washed with water, and freeze dried.

(2) This system will separate equally well the products of digestion of insulin derivatives that have —COOH as well as —NH_2 protection.

(3) LH-20 suffers from the disadvantage that it begins to exclude, and thus fails to separate, molecules that are much larger than those separated here. Sephadex LH-60 has recently become available and should prove useful in such instances. The nature of the solvent, which need not be semi-aqueous as in the Method above, affects the range of separation.

Galpin *et al.* (1976) reported that the gel-filtration medium Enzacryl is suitable for gel filtration in organic solvents and shows a cut-off at a much higher molecular weight, but more recently they have begun to prefer Sephadex LH-60 (Galpin *et al.*, 1978, 1979).

CHAPTER 4

Protection after cleavage

If the operations discussed in Chapters 2 and 3 have been successful, we are now in possession of a set of partially protected fragments. They require some further protection before they can be used in the reassembly of the protein, but these operations are often quite simple. Also, in sharp contrast with conventional synthesis, the effort in preparing them does not increase at all sharply with chain length.

Amino protection

Of the —NH_2 groups of the fragments, only those on the α-carbon atom still require protection. Such protection is necessary so that the fragments can be used in controlled couplings, but even if the fragment is destined to become an amino component, temporary protection of the α-NH_2 is required if it is first going to be esterified by a diazoalkane (p. 71). A group must be chosen that is compatible with those on the side-chains. The Boc- group is suitable when the side-chain protection involves the acetimidyl, Cbz-, phthaloyl, acylmethionyl, or trifluoroacetyl groups. It is also compatible with the maleyl group—Rees and Offord (1976a) showed that the maleyl group, though deprotected in aqueous acid, is stable to anhydrous trifluoroacetic acid. Example (4.1) describes a typical α-protection. Example (4.2) describes the use of an acid-stable group, which would then be compatible with acid-labile protection of the side-chains. There is little reason to doubt that all the methods of —NH_2 protection discussed in Chapter 2 can be adapted for use on fragments: the major changes would have to be made in the isolation of the protected peptide. The smaller the peptide the more difficult it would be to separate it from unwanted small molecules such as excess reagent and its by-products.

Carboxyl protection

Example (2.10) indicated a method for the acid-catalysed methylation of a protein. The method is equally applicable to peptides.

In contrast, the more popular methods for protection of —COOH that were described in Chapter 1 were never designed to be used in the presence of a peptide bond. They are usually too severe to be used on peptides without damage. For instance, Backer and Offord (1968) found that, using benzyl alcohol, the benzyl group could only be applied to peptides under conditions so vigorous that some cleavage of the peptide chain occurred. On the other hand, benzylation by phenyldiazomethane proceeded smoothly in semi-aqueous solutions at an apparent pH of between 4.5 and 5.0.

$$C_6H_5-CH=\overset{+}{N}=\overset{-}{N} + H^+ \rightleftharpoons C_6H_5-CH_2^+ + N_2$$

$$C_6H_5-CH_2^+ + R-COO^- \rightleftharpoons R-CO-O-CH_2-C_6H_5 \qquad (4.1)$$

The only noticeable side-reactions with ordinary peptides were a considerable degree of alkylation of unprotected imidazole groups and a very slight degree of alkylation of free $-NH_2$ groups. Failure to remove the by-products of the reagent can lead to extensive damage to tryptophan residues on subsequent acid deprotection (Offord *et al.*, 1976). Harris (1977) found that esterification by a diazoalkane irreversibly altered the spectroscopic properties of a haem peptide while acid-catalysed methylation did not.

Example (4.3) gives generally useful conditions for esterification with diazo-alkanes.

Example (4.4) describes the preparation of phenyldiazomethane. The benzyl group occupies a useful place in peptide synthesis, but, as pointed out in Chapter 1, it is not particularly easy to remove subsequently. The search for reagents that introduce more readily acid-labile groups led Hayward and Offord (1971) to propose the use of 1-diazo-2,2-dimethylpropane.

$$CH_3-C(CH_3)_2-CH=\overset{+}{N}=\overset{-}{N} + H^+ \rightleftharpoons CH_3-C(CH_3)_2-CH_2^+ + N_2$$

$$CH_3-C(CH_3)_2-CH_2^+ \rightleftharpoons CH_3-\overset{+}{C}(CH_3)-CH_2-CH_3 \qquad (4.2)$$

The key feature of Reaction (4.2) is the rearrangement of the secondary carbonium ion to a tertiary one. If this process can go to completion before the esterification step, we will obtain, not the 2,2-dimethylpropyl ester, which would not be at all acid-labile, but the 1,1-dimethylethyl ester. This is the next higher homologue of the t-butyl ester, and its acid lability should be very similar or even a little greater. (A diazoalkane that applies the t-butyl group itself is clearly an impossibility.)

Example (4.5) describes the preparation and use of the 1-diazo-2,2-dimethyl-propane. Its principal advantage is that it applies a group that is resistant to hydrazine (Hayward and Offord, 1971). Therefore, in contrast to most other diazoalkanes its use is compatible with coupling by the azide route. It suffers the disadvantage that the reagent is rather unstable in the semi-aqueous, mildly acidic solutions that are best for peptide esterifications with diazoalkanes. Very large excesses of reagent are required unless one decreases the quantity of water in the solvent. A balance has to be struck between the need to minimize the quantity of water present and the need to keep up the dielectric constant of the medium, to facilitate the rearrangement of the carbonium ion. The Notes to Example (4.4) indicate some of the solvent mixtures that have been found satisfactory (and which can also be used with other diazoalkanes).

The Example shows that to control the reaction we add, simultaneously, two solutions of different polarity. This makes it quite difficult to keep the dielectric constant of the reaction mixture between specified limits. Even under conditions that are normally satisfactory an accidental excursion toward too low a value for the dielectric constant will lead to the formation of some of the primary ester, which is not acid labile. Offord *et al.* (1976) have found as little as 94% of a monocarboxylic ester to be acid labile when prepared under conditions that had given no detectable quantity of the incorrect product on a previous occasion. This problem, though tiresome, is not an insuperable drawback when only one or two —COOH groups are to be protected. The yield of acid-labile product will be given by the usual sort of $(q/100)^n \times 100$ relationship. If we take our example of 94% deprotection per —COOH group, a protein such as insulin, with 6 —COOH groups per molecule will have $(0.94)^6 = 69\%$ of its molecules capable of being completely deprotected. Nonetheless, where the —COOH protection does not have to resist hydrazinolysis, it would clearly be better to seek a reagent that operates on a simpler principle.

Diphenyldiazomethane has been used in conventional peptide synthesis (e.g. Aboderin *et al.*, 1965; Stelakatos *et al.*, 1966) and gives an acid-labile ester. It appears to modify —COOH groups in natural fragments but it does so very slowly (see Note 4 to Example (4.3)), and the group introduced is so bulky and apolar that the solubility of the products can cause difficulty. Offord *et al.* (1976) have proposed instead the use of *p*-methoxyphenyldiazomethane, which produces *p*-methoxybenzyl esters. This group has been employed in conventional peptide synthesis (e.g. Stelakatos and Argyropoulos, 1971) although it was not applied via a diazoalkane. As the known stability of the *p*-methoxybenzyl carbonium ion would suggest, the ester is considerably more acid labile than are benzyl esters. Rees and Offord (1976a) report that it is more readily deprotected from the side-chains of glutamic-acid residues than from those of aspartic-acid residues but that in any case the required times of exposure to trifluoroacetic acid are not excessive. Example (4.6) describes the preparation and use of the diazoalkane.

It has been said several times that loss of ionizable side-chains on protection leads to problems of insolubility, and also that having protecting reagents that are chemically inert but ionizable may facilitate a sterically assisted coupling between large fragments. Offord *et al.* (1976) have given a preliminary report on a diazoalkane, 4-methoxy-3-sulphophenyldiazomethane, which produces esters that have not lost the negative charge. Example (4.7) describes its preparation and use. It is not so easy to use as the other aromatic diazoalkanes but is probably easier to make than 1-diazo-2,2-dimethylpropane. Its real usefulness in semisynthesis has yet to be evaluated.

Diazomethane itself can be used for methylation if the acid-catalysed method is unacceptable for any reason. It can be used under the ordinary, semi-aqueous conditions of Example (4.3). It has also been tried on natural fragments in anhydrous solution (Offord *et al.*, 1976). Alkylation of unprotected imidazole and $—NH_2$ groups was found to be particularly serious in such solvents. Anhydrous solvents should in principle be equally suitable for the esterification of $—NH_2$-protected peptides by diazoalkanes, but there seem to be no reports of their having been so used.

Diazoalkanes must be treated with respect, but their acute toxicity and risk of explosion are not excessive. (The latent effects of chronic, low-level exposure to the diazoalkanes and their precursors remain to be seen.) They are adequate for the protection of natural fragments, but they are not perfect. Workers in this Laboratory have frequently sought alternatives to diazoalkanes, but, with the exception of the acid-catalysed methylation by methanol, none have yet been found that were as simple and reliable. This is not to say that suitable alternatives will not be found in the future.

Glass and Pelzig (1977) have described in detail their approach to —COOH protection by enzyme-labile substituents. The principle of the method is to couple an arginine ester to the —COOH groups in the peptide or protein. When the time comes to deprotect, the ester group is removed and then, by means of carboxypeptidase B, the arginine. Since this exopeptidase will only attack basic residues, the method is in theory applicable to all molecules that do not have lysine or arginine at the carboxyl terminus of the final product. This approach is certainly worthy of further development, particularly in the light of the generally unsatisfactory nature of the methods available for —COOH protection. One potential advantage of the method is that the protection takes place through the formation of an amide rather than an ester bond, which is generally easier to do, particularly if (as is true of side-chain protection) racemization is not an issue. Glass and Pelzig use 1-hydroxybenzotriazole and carbodiimide to couple on the arginine ester. The guanido group will help to keep the product water soluble, although with a reversal of charge which will presumably count against the possibility of sterically assisted coupling of fragments (p. 29). In their model studies, Glass and Pelzig have used the methyl ester of arginine, which they deprotect with trypsin. In actual semi-syntheses, the product will be likely to have trypsin-labile peptide bonds and they will have to use either an ester group that is removed in a different way or (as is quite possible) conditions that favour only the esterolytic and not the proteolytic activity of trypsin.

The removal of all the esters described here is discussed in Chapter 7.

The selective protection of α-COOH groups

It was said in Chapter 2 that the question of obtaining differential protection between α- and side-chain —COOH groups was best left until after the protein had been cleaved and the fragments isolated. Side-chain protection prior to cleavage is a satisfactory means of obtaining differential $—NH_2$ protection, largely because the effect on solubility of the protecting groups used is not too serious. The groups chosen either enhance or at least do not prohibitively affect solubility in the media used for proteolysis and for the subsequent separation of fragments. On the other hand, when we protect —COOH groups by esterification, all but the 4-methoxy-3-sulphobenzyl ester and the methyl ester tend to depress solubility severely. Of these two, the sulphonated reagent has as yet been little used, and the methyl group is disliked because of the basic conditions needed for its removal, and because of fears of $\alpha \rightarrow \omega$ transpeptidation (p. 15).

The time will probably come when we have methods for —COOH protection prior to cleavage which we can apply with confidence. Until then, we have some quite satisfactory methods for differentiation between α- and side-chain —COOH after cleavage. The nature of these methods depends on whether we are dealing with a tryptic or a CNBr fragment.

Selective α-deprotection of tryptic fragments

We rely on the fact that trypsin is an α-COOH esterase as well as a protease (Schwert *et al.*, 1948). If we esterify the fragment, both types of —COOH group will be protected.

$$H_2N\text{—}\underset{|}{\overset{COOH}{}}\text{——}\underset{|}{\overset{COOH}{}}\text{——}COOH \longrightarrow H_2N\text{—}\underset{|}{\overset{CO—O—R}{}}\text{——}\underset{|}{\overset{CO—O—R}{}}\text{——}CO—O—R \qquad (4.3)$$

If the fragment was obtained from a limited tryptic digest (p. 53) it will have arginine at the carboxyl terminus. A second application of trypsin will selectively liberate the α-COOH group of the arginine (Backer and Offord, 1968; Offord, 1969a; Offord *et al.*, 1976).

$$H_2N\text{—}\underset{|}{\overset{CO—O—R}{}}\text{——}\underset{|}{\overset{CO—O—R}{}}\text{——}CO—O—R \longrightarrow H_2N\text{—}\underset{|}{\overset{CO—O—R}{}}\text{——}\underset{|}{\overset{CO—O—R}{}}\text{——}COOH \qquad (4.4)$$

Example (4.8) gives the conditions. It will be observed that, by following the reaction in the autotitrator, we obtain a quantitative estimate of the amount of —COOH liberated.

Only the fragment from the carboxyl-terminus of the protein, which will not usually end in arginine, is unsuitable for Reaction (4.4). However, this fragment will usually only be needed as an amino component, and so Reaction (4.3) alone will suffice.

Trypsin appears to be a more active esterase (in terms of moles of substrate cleaved per unit time) than it is a protease. Possibly because of this, the restriction of its esterolytic action to derivatives of lysine and arginine is less sharply marked (Kloss and Schroder, 1964). Other α-COOH esters of amino acids and peptides can be cleaved but not, so far as is known, any side-chain esters and so this finding extends, rather than restricts, the use of Reaction (4.4).

What has been said of the treatment of tryptic fragments can in theory be applied to the fragments produced by any other protease that also has the appropriate esterase activity.

Selective protection of CNBr fragments

Offord (1972) showed that CNBr fragments can be differentially protected while the homoserine at the carboxyl-terminus is still in the lactone form in which it is liberated by Reaction (3.5). First the free —COOH groups are esterified

$$H_2N\text{—}(COOH)\text{—}(COOH)\text{—}NH\text{—}\underset{\text{lactone: } H_2C\text{—}C(H_2)\text{—}O\text{—}C(=O)}{CH}\text{—}C \longrightarrow H_2N\text{—}(CO\text{—}O\text{—}R)\text{—}(CO\text{—}O\text{—}R)\text{—}NH\text{—}\underset{\text{lactone: } H_2C\text{—}C(H_2)\text{—}O\text{—}C(=O)}{CH}\text{—}C \quad (4.5)$$

and then the lactone is opened.

$$H_2N\text{—}(CO\text{—}O\text{—}R)\text{—}(CO\text{—}O\text{—}R)\text{—}NH\text{—}\underset{\text{lactone: } H_2C\text{—}C(H_2)\text{—}O\text{—}C(=O)}{CH}\text{—}C \longrightarrow H_2N\text{—}(CO\text{—}O\text{—}R)\text{—}(CO\text{—}O\text{—}R)\text{—}NH\text{—}\underset{CH_2\text{—}CH_2\text{—}OH}{CH}\text{—}C(=O)\text{—}OH \quad (4.6)$$

There is no danger of the premature opening of the lactone if Reaction (4.5) is carried out by acid-catalysed methylation—the lactone ring appears to escape methanolysis (Wallace, 1976). The weakly acidic, semi-aqueous conditions for an esterification by a diazoalkane (e.g. Example (4.3)) in theory offer a chance for the lactone to open, and its —COOH group to be esterified, but in practice the opening is far too slow for it to have any detectable effect on the nature of the product. Reaction (4.6) is a saponification, and although the lactone is unusually reactive, very mildly basic conditions have to be employed if the side-chain ester groups are not to be partially lost while the ring is being opened (Wallace, 1976; Wallace and Offord, 1979). Example (4.9) is typical.

We shall see in Chapter 6 that there are other ways in which we can make use of the special reactivity of the lactone to carry out controlled couplings. However, Reactions (4.5) and (4.6) continue to provide the best ways of preparing a CNBr fragment for the removal of the homoserine residue in order subsequently to restore the methionine (e.g. Example (5.3)).

Imidazole protection

We have seen that the main reason for protecting the imidazole group is to prevent N^{im}-alkylation by diazoalkanes and that protection can be applied before or after cleavage. If the decision is taken to protect, the only reason for delaying until after cleavage would be to have the fragment as polar as possible during the separation of the fragments. Example (4.10) describes the protection of the imidazole of a N^{α}-Boc- fragment by one of the 1-carbamido-2,2,2-trifluoroethane type of reagents mentioned in Chapter 1.

The other imidazole-protecting groups in Table 1.6 can also be applied to biological fragments. If the Boc- group is applied to the α-NH_2 group under the conditions given in Example (4.1), the imidazole group will almost certainly be protected. Oddly, the group then constitutes a differential protection. The N^{im}-Boc- group is stable to 2 M HCl in dioxan (Fridkin and Goren, 1971), while the α-Boc- is not; the N^{α}-Boc- group is stable to 2 M acetic acid while the N^{im}-Boc- is not. However, the N^{im}-Boc- group is also very readily removed by nucleophilic displacement (Schnabel *et al.*, 1971), and it is insufficiently stable to some of the conditions likely to be encountered in semisynthesis (e.g. those of Example (4.3)). It is best to remove it as described in Note 2 to Example (4.1) and replace it, if desired, by a more stable group as described in Example (4.10). Rees (1976) found that a coupling in which the amino-terminal residue of the amino component was histidine went much faster when the N^{im} protection was Boc- than when it was Boc—NH—CH(CF_3)-.

EXAMPLES

Example (4.1). The N^{α}-protection of an N^{ε}-acetimidylated 69-residue peptide

Source. Milman (1974).

Process involved. Reaction (1.12) in which R = $(CH_3)_3C$—.

Method

1 μmol of the 69-residue peptide (obtained from peak 1 in Example (3.7), Method A) was dissolved in 0.5 ml trifluoroacetic acid. The acid was removed *in vacuo* after 1.5 h at 20 °C. The peptide was then dissolved in dimethylsulphoxide (0.16 ml) and brought to an externally indicated pH (p. 217) of 8.5 with redist. *N*-ethylmorpholine. t-Butyloxycarbonyl azide (14.3 μl, 100 μmol) was added and the externally indicated pH adjusted to about 9 (with approximately 10 μl of *N*-ethylmorpholine). The mixture was allowed to stand at 20 °C. Paper-chromatographic tests on 4 μl samples of the mixture showed that the majority of the peptide was protected after 10 h, but that the reaction was not complete until 46 h had elapsed. The mixture was dried down *in vacuo* (approximately 0.1 Torr—13 Pa) and washed with ether.

Notes

(1) The peptide, which has homoserine at the carboxyl terminus, was obtained in the open form (see Note 6 to Example (3.7)). The treatment with trifluoroacetic acid was intended to convert it into the lactone form. Such a conversion is necessary to ensure subsequent differentiation between side-chain and α-COOH groups (p. 122). It cannot be done with trifluoroacetic acid once the acid-labile Boc- group has been applied. It is therefore fortunate that the prolonged exposure to base does not reopen the lactone, presumably because the reaction medium is anhydrous (contrast Example (4.9)). Note 1 to Example (5.3) explains why it might in practice be possible to omit the trifluoroacetic acid treatment for some applications, but if the peptide is not otherwise harmed by it, it may be best to carry out the treatment and be sure of complete conversion in all cases.

(2) If the peptide contains histidine, the conditions given above are in addition likely to lead, to a greater or lesser extent, to the Boc- protection of the imidazole group. For the reason given on p. 74, the N^{im}-Boc- group is unsuitable for semisynthetic use, and it is best to remove it. This can be done by dissolving or suspending the product (10 μmol/ml) in 2 M acetic acid for 1 h at 0 °C and then freeze drying. N^{α}-Boc- groups are stable to this treatment. See also Example (4.10).

(3) As remarked previously, di-t-butyloxydicarbonate can be used instead of the azide reagent. The following preparation (carried out on a small peptide: C. J. A. Wallace, unpublished work) is typical.

Peptide (10 μmol) was dissolved in 2 ml 0.1 M *N*-ethylmorpholine buffer (adjusted to pH 8.0 with strong acetic acid). To this was added the dicarbonate (21.8 mg, 100 μmol) and the mixture was stirred vigorously for 1 h at 20 °C. A second, equal quantity of reagent was then added and stirring continued for another hour. The mixture was then freeze dried.

Example (4.2). The N^{α}-protection of a tryptic peptide by the Cbz- group

Source. Offord (1969a).

Process involved. Reaction (1.12a) in which R = $C_6H_5CH_2$—.

Method

Peptide (10 μmol) was dissolved in 1 ml ethanol/water (1 : 1, v/v) at 0 °C. The solution was stirred vigorously throughout all subsequent operations. Powdered MgO (100 mg) was added to the solution. Benzyloxycarbonyl chloride (10 μl, 70 μmol) was added in 2 μl lots at half-hour intervals while the solution was kept in the ice bath. More reagent (12 μl, 84 μmol) was then added, and the solution allowed to warm up to room temperature overnight. If electrophoresis of a 4 μl sample of suspension showed incomplete protection (see Example (A.6)) more reagent was added. Normally, protection was complete and the MgO was spun down, extracted once with 1 ml ethanol/water (1 : 1, v/v) and once with pyridine/water (1 : 1, v/v). The washings were united with the supernatant and dried down. The residue was extracted with ether. The peptide was normally in the ether-insoluble residue and it could be confirmed by end-group analysis (Example (A.2)) that it was fully protected.

Notes

(1) As in previous Examples, we find that a much greater excess of reagent is called for in reactions with peptides and proteins than is normal when protecting amino acids. An amino acid would normally have been completely protected after the addition of the first few μl of reagent.

(2) Because of the large excess of reagent, a considerable quantity of base is likely to be needed to deal with the products of its decomposition. The present Example uses MgO because it was recommended by Baer and Eckstein (1962) as unlikely to promote benzyloxycarbonylation of side-chain —OH groups and because, as a solid, it represents a large, hidden reserve of mild basicity. It was used for the latter reason by Borrás and Offord (1970a, b) to control the *N*-trifluoroacetylation of insulin by the aminolysis of phenyl trifluoroacetate, but other bases, such as *N*-ethylmorpholine, may be just as effective in this respect (cf. Note 1 to Example (2.1) and Note 3 to Example (6.6)).

(3) The extraction of the MgO is usually very effective in removing adsorbed peptides and final yields are normally good.

Example (4.3). The esterification of peptides by diazoalkanes in semi-aqueous solution

Source. Offord *et al.* (1976).

Process involved. Reaction (4.1). R can be CH_3—, $CH_3CH_2C(CH_3)_2$—, $C_6H_5CH_2$—, $(C_6H_5)_2CH$—, or *p*-$CH_3OC_6H_4CH_2$—. The preparation of the various diazoalkanes is described in subsequent Examples.

Methods

The peptide was dissolved (up to 0.1 mol) either in water or (if insoluble in water by virtue of other protecting groups) in the organic component of the proposed mixture. (See Note 1 for the possible range of compositions.) The aqueous and organic components were mixed, with cooling if the process proved to be exothermic. The peptide solution was then transferred to an autotitrator vessel, which was surrounded by ice, and the apparent pH (glass electrode, standardized in aqueous buffer at 0 °C) brought to 4.5 with aqueous HCl (usually between 0.01 and 0.1 M). The solution was stirred vigorously at all times.

The ethereal solution of diazoalkane (which was kept near the autotitrator in a receptacle cooled in isopropanol/solid CO_2) was then added drop by drop. The pH tended to rise after each addition, and matters were arranged so that the autotitrator was able to prevent any excursion above pH 5. At first the tendency to rise was very sharp, and the rate of evolution of nitrogen and the loss of colour of the reagent were very rapid. As the addition progressed these changes slowed, and the acid uptake became less extensive, settling down to values that presumably reflected the decomposition of the reagent rather than its reaction with —COOH groups. Esterification (as judged by subsequent electrophoresis of the product) was usually complete when a 30- to 70-fold excess of reagent over —COOH groups had been added.

After the reaction was judged to be complete, the mixture was taken briefly to pH 3.5 to decompose any traces of diazoalkane and then to pH 7 (with 0.1 M pyridine in water, or any other convenient base). The ether was removed on the rotary evaporator. A copious precipitate, believed to be of by-products derived from the diazoalkane, was usually formed during this operation. This precipitate was removed by centrifugation, and after testing for peptide material, discarded. The remaining solution was then taken to dryness and the peptide material extracted from most of the remaining by-products into pyridine/water (1:1, v/v).

Notes

(1) It is possible to select as the organic solvent methanol, dimethylformamide, or ethanol, as required by the solubility of the starting material. Dimethylsulphoxide is to be avoided (although it is a very good solvent) because its use appears to lead to a heterogeneous product—notably in peptides that have aromatic residues. The water content is usually set at 15% (v/v) because more peptides appear to be soluble at this sort of ratio than at any other. At this ratio there was not so much water as to bring about the immediate formation of a second layer on adding the ethereal solution of diazoalkane, but there

was enough to permit pH control with a glass electrode. If a great deal of ethereal solution had to be added, the solution separated into two phases. If this occurred some more of the original solvent mixture was added to restore the system to a single phase. Rees and Offord (1976a) have occasionally used mixtures in which the quantity of water ranged from 10% to 50% by volume.

(2) Even if the peptide was initially insoluble in the solvent system it usually went into solution during the reaction.

(3) The pH was a compromise value. The rate of decomposition of most diazoalkanes is prohibitively fast below pH 4 and it is desirable to keep the pH below 5 to minimize the possibility of reactions occurring with functional groups other than —COOH.

(4) Diphenyldiazomethane reacted so slowly as to be of very limited usefulness. It was also much slower to decompose. At pH 4 (glass electrode, 0 °C) the apparent first-order rate constant of the acid uptake that accompanied its decomposition in water/dimethylformamide 15 : 85 in the absence of substrate was $4.18 \times 10^{-4}\ s^{-1}$ (Offord *et al.*, 1976). The corresponding value at pH 3 was $1.75 \times 10^{-3}\ s^{-1}$. By contrast the decomposition of the aliphatic diazoalkanes was practically instantaneous at pH 3, and still too fast for the autotitrator to follow at pH 4.

(5) Trial experiments have occasionally shown larger excesses of diazoalkane to be necessary. The quantity of diazoalkane in the stock solution was assessed approximately by the extinction data in Note 4 to Example (4.6).

(6) An alternative method for isolating the peptide ester from the reaction mixture is, after the removal of the precipitate, to apply the solution for gel filtration on Sephadex LH-20. Dimethylformamide/water (17 : 3, v/v) or some other appropriate solvent mixture can be used. The size of column described in Example (3.8) for the separation of tryptic digests is suitable, but it is prudent to keep a separate column specifically for peptide esters. Peptide esters are very sensitive to traces of trypsin (p. 73).

Gel filtration is a more efficient means of removing the last traces of the by-products of the reaction than is the extraction described in the Method. Gel filtration should therefore be used if the peptide is subsequently to undergo strong acid treatment, since the by-products appear to promote damage to aromatic side-chains under such conditions. It is sometimes essential to use the extraction method when the peptide contains tryptophan, since this residue can lead to the late emergence of the peptide from the column, with consequent overlap with the by-products.

(7) It is possible to esterify in non-aqueous solutions, as is more usual in organic chemistry. When necessary, the substrate was first converted into the carboxylic-acid, as opposed to the carboxylate, form by being dissolved or suspended in cold water which was then adjusted to pH 2.5 with 0.01 M HCl. The liquid was freeze dried at once and redissolved in the desired solvent. Offord *et al.* (1976) reported that numerous experiments with small substrates (up to 6 amino-acid residues in length) showed that, in general, a 3-fold excess of diazoalkane was sufficient to bring about complete esterification in a few

minutes at 0 °C. Diphenyldiazomethane was exceptional in that it was extremely slow to react. Even a large excess often failed to esterify completely quite small substrates after 24 h.

(8) Because of the danger of *N*-alkylation, it is essential to protect imidazole groups before esterification (Example (4.10)) and desirable to protect $-NH_2$ groups by one of the methods described previously. *N*-Alkylation is a much more severe problem in the absence of water.

(9) The specific α-deprotection of the peptide esters is described in Example (4.8). The conditions for the general deprotection of all —COOH groups are discussed in Chapter 7.

(10) All diazoalkanes and their precursors should be treated as though they were highly toxic.

Example (4.4). The preparation of phenyldiazomethane

Source. Rees (1974), based on Sarin and Fasman (1964).

Process involved.

benzil $\xrightarrow{N_2H_4}$ benzil monohydrazone $\xrightarrow{[O]}$

azibenzil $\xrightarrow{OH^-}$ benzoate + phenyldiazomethane (4.7)

Method

Hydrazine sulphate (52 g) and sodium acetate (110 g) were dissolved together in 250 ml of water and boiled for 5 min. The solution was then cooled to 50 °C and 224 ml of methanol added. There was a white precipitate of Na_2SO_4. The precipitate was filtered off and washed with a little ethanol. The combined filtrates were heated to 60 °C. To this was added a warm solution of benzil (50 g) in methanol (74 ml). The mixture was refluxed for 30 min and then allowed to cool to room temperature. The pale-yellow precipitate of benzil monohydrazone was filtered off, washed well with ether (the yellow colour was thereby removed), dried and stored in a light-proof bottle. Yield 42.5 g, m.p. 149–152 °C.

The hydrazone (18 g) was ground in a mortar and pestle with yellow HgO (36 g) and anhydrous Na_2SO_4 (9 g) until the components were thoroughly mixed. The powder was then transferred to a 500 ml round-bottomed flask containing 110 ml ether. A saturated solution of KOH in ethanol (4 ml) was then added and the mixture stirred vigorously for 20 min. The inorganic materials were filtered off and washed with ether. The combined filtrates, which were red in colour, were taken to dryness at 30 °C in a rotary evaporator situated behind a protective screen. The azibenzil was recrystallized from Na-dried ether at −20 °C. Yield 11 g, m.p. 78–79 °C.

Azibenzil (5.5 g) was dissolved in 125 ml of Na-dried ether in a 500 ml flask. To this was added a solution made up by dissolving NaOH (8 g) in 15 ml of H_2O, and then diluting with 105 ml of methanol. The whole mixture was stirred in the dark at 20–25 °C for 11 h. The red, single-phase solution was then shaken in a separating funnel with 100 ml of 10% (w/v) NaOH. The ethereal layer was washed with 6 × 25 ml of 10% NaOH and with 25 ml lots of water until the washings were neutral. The ethereal solution of diazoalkane (approximately 0.2 M) was dried over anhydrous Na_2SO_4 and stored in a number of stoppered vials at −20 °C. It was stable for some weeks.

Notes

(1) Phenyldiazomethane is probably much less likely to explode than diazomethane.

(2) There is very little build-up of N_2 in the storage vials.

(3) Solid diphenyldiazomethane can be prepared by the method of Miller (1959). Benzophenone hydrazone (2.3 g), anhydrous Na_2SO_4 (3.0 g), and yellow HgO (7.0 g) were mixed with 40 ml of Na-dried ether in a screw-top glass bottle. 1 ml of a freshly made, saturated solution of KOH in ethanol was added and the mixture was shaken mechanically in the fume hood. The mixture was then filtered and rotary-evaporated to an oil. 30 ml of petroleum ether (40–60 °C) were added, the mixture filtered once again and evaporated to dryness. The deep-red oil was then cooled in an isopropanol/solid CO_2 bath. Needle-like crystals deposited as the material warmed up. The materials were stable if stored at 2 °C.

(4) For the assay of these materials see Note 4 to Example (4.6).

Example (4.5). The preparation of 1-diazo-2,2-dimethylpropane

Source. Offord *et al.* (1976), based on Gerber (1952).

Process involved.

$$\underset{\text{amine}}{CH_3-\overset{\displaystyle CH_3}{\underset{\displaystyle CH_3}{\overset{|}{\underset{|}{C}}}}-CH_2NH_2} + \underset{\text{ethyl chloroformate}}{Cl-CO-O-CH_2-CH_3} \longrightarrow$$

$$\underset{\text{urethan}}{CH_3-\overset{\displaystyle CH_3}{\underset{\displaystyle CH_3}{\overset{|}{\underset{|}{C}}}}-CH_2-NH-CO-O-CH_2-CH_3} \xrightarrow{NO_2}$$

$$\underset{\text{N-nitrosourethan}}{CH_3-\underset{CH_3}{\overset{CH_3}{C}}-CH_2-\underset{NO}{N}-CO-O-CH_2-CH_3} \xrightarrow{\text{base}} \underset{\text{diazoalkane}}{CH_3-\underset{CH_3}{\overset{CH_3}{C}}-CH=\overset{+}{N}=\overset{-}{N}} \qquad (4.8)$$

Method

1-Amino-2-2idimethylpropane. 25 g of $LiAlH_4$ were suspended in 500 ml of Na-dried ether. t-Butyl nitrile (41 g) was dissolved in 125 ml of Na-dried ether and added to the $LiAlH_4$ solution at such a rate that it refluxed gently. The mixture was allowed to stand at room temperature for 1 h, after which 20 ml of water were slowly added, followed by NaOH solution (3 g in 20 ml water) and then a further 75 ml water. The ethereal layer was collected, and the aqueous layer was washed with a further 150 ml of ether. The combined ether layers were distilled through a Vigreux column. 25.8 g (87 % theoretical yield) of a colourless liquid b.p. 79–82 °C were collected. $n_D^{20} = 1.402$.

2,2-Dimethylpropan-1-urethan. 40 g of the amine were dissolved in 200 ml of Na-dried ether and cooled to 5 °C. Freshly distilled ethyl chloroformate (64 g) was added slowly with vigorous stirring. The temperature was kept below 10 °C. A white precipitate began to form at once. Half-way through the addition of chloroformate, a parallel addition of 36 ml water containing 23.2 g of NaOH was begun. After the addition of all the reagents was complete, 100 ml of water was added. The ethereal layer was removed and extracted with 100 ml water. The combined aqueous layers were extracted twice with 50 ml of ether. The combined ethereal layers were dried for 1 h over Na_2SO_4. After distillation of the ether at atmospheric pressure, the urethan was distilled as a colourless oil, b.p. 99–100 °C at 24 mmHg. The yield was 49.6 g (68 %). The infrared spectrum showed the expected absorption at 3300 cm^{-1} and 1700 cm^{-1}.

N-Nitroso-2,2-dimethyl-1-urethan. 15 g of the urethan were dissolved in 75 ml of ether and added to a solution of $NaNO_2$ (30 g) in 45 ml water. The mixture was cooled in a round-bottomed flask to 5 °C. 28 ml HNO_3 and 28 ml of ice-cold water were mixed and added over the course of 1 h through a separating funnel that had a stem extended to below the interface between the aqueous and the ethereal layer. The acid must *not* enter the ether layer. The reaction mixture was allowed to stand for 20 h at room temperature. The ethereal layer, which was yellow, was treated as a potential carcinogen. It was separated and extracted with 5 changes of 76 ml each of water, and then with 75 ml lots of 1 M KH_2PO_4 adjusted to pH 7 with 1 M NaOH, until the aqueous layer had a pH of 7 (4–5 changes were usually needed). The aqueous layers were extracted in turn with another, small sample of ether. The combined ethereal layers were kept over solid, dry $NaHCO_3$ for 1 h and then over Na_2SO_4 overnight. The solution was not characterized because of the possible danger in handling it.

1-Diazo-2,2-dimethylpropane. A vacuum-distillation apparatus was set up with provision for the system to be purged by N_2 via an anti-bumping bleed of somewhat greater than normal internal diameter (approximately 0.2 mm). The flow of N_2 was controlled both for bleeding and for anti-bumping by a screw clip on the rubber feeder tube. Two cold-finger tubes in series, cooled by solid CO_2/isopropanol, were used to collect distillate. Provision was made for warming the distillation flask to 30 °C, for magnetic stirring of its contents, and for the addition of liquid reagents while the system was under vacuum. The ethereal solution of nitroso compound (from 7.5 g of urethan) was introduced into the distillation flask together with a heavy magnetic stirring bar and the ether removed under water-pump vacuum with air bleed. The system was then brought to atmospheric pressure, 20 ml of ether added, and the distillation repeated. This cycle was repeated until the distillate (which was removed from the collector traps each time) no longer imparted an acid reaction to distilled water. 10 ml of fresh ether were then placed in each trap and the system connected to an oil pump capable of an ultimate vacuum of 13 Pa (0.1 Torr) or less and connected to a supply of oxygen-free N_2 via the bleed. The pump was protected from the system by at least one additional cold trap. The system was purged with N_2 and, with vigorous stirring, 12 ml of a solution of 1.25 g of Na in 25 ml of ethanediol was added for every 7.5 g of urethan that had been nitrosylated. The reaction mixture went bright orange. Vigorous stirring of the viscous mixture was continued throughout. The system was then taken to the lowest possible pressure. In order to achieve such a low pressure it was important to restrict the N_2 supply to the bleed to the absolute minimum and to lubricate any suspect ground-glass joints by external application of ethanediol. With these precautions the diazoalkane distilled over within a few minutes, and was seen to form a strongly orange solution in the first trap and a weakly orange solution in the second trap. The mixture in the reaction vessel (which could be warmed to 30 °C to assist the distillation without apparent harm) had lost nearly all its colour by the end of the distillation.

Solutions of diazoalkane could be prepared in this way up to a concentration of approximately 0.2 M with a yield (based on urethan) of about 2%. The yield is assessed as described in Note 4 to Example (4.6). They were stored in a number of small vials. If the last traces of acid were removed as described above the solutions were stable for a few weeks at −20 °C.

Notes

(1) The nitrosourethan is likely to be carcinogenic. The danger of the diazoalkane exploding is likely to be less than that experienced with diazomethane.

(2) The Method is a minor adaptation of that of Gerber—the changes are mainly in the distillation of the product and were found by Offord *et al.* (1976) to increase the yield of diazoalkane.

(3) The best solvents for promoting the rearrangement to the tertiary carbonium ion were ethanol/water (17 : 3, v/v) and dimethylformamide/water (17 : 3, v/v). Hayward and Offord (1971) also used ethanol/water (197 : 3, v/v) apparently without undesirable consequences.

(4) C. F. Hayward (private communication) prepared *N*-nitroso-2,2-dimethylpropan-1-toluenesulphonamide (m.p. (decomp.) 55–56 °C, with the expected N—NO stretching absorption at 1520–1540 cm^{-1}). This compound ought to be a less hazardous precursor of the diazoalkane, but repeated attempts using the method above failed to give any product whatsoever. However, the method with the nitrosourethan is so dependent on the quality of the vacuum within the apparatus (as opposed to its quality at the point of measurement, near the intake of the pump) that it is possible that some minor change in the procedure might lead to success with the nitrosotoluenesulphonamide.

(5) Diazomethane is best prepared by an analogous but much simpler method. The *N*-nitroso precursors are commercially available. The procedure of de Boer and Backer (1963) is typical. 2.14 g of *N*-methyl-*N*-nitrosotoluenesulphonamide were dissolved in 30 ml of diethyl ether in a distillation apparatus made with polished joints. A solution of 0.4 g KOH in 10 ml of 96% ethanol was then added. The flask was gently warmed in a water bath (60 °C) and the ethereal solution of diazomethane collected in a receiver cooled in ice. Yield about 0.3 g diazomethane, assayed as described in Note 4 to Example (4.6).

Example (4.6). The preparation of *p*-methoxyphenyldiazomethane

Source. D. A. Wightman (unpublished), based on Curtius (1912) and Overberger *et al.* (1965).

Process involved.

$$CH_3{-}O{-}C_6H_4{-}CHO \xrightarrow[\text{(aqueous)}]{N_2H_4}$$

p-methoxybenzaldehyde

$$CH_3{-}O{-}C_6H_4{-}CH{=}N{-}N{=}CH{-}C_6H_4{-}O{-}CH_3 \xrightarrow[\text{(anhydrous)}]{N_2H_4}$$

anisalazine

$$CH_3{-}O{-}C_6H_4{-}CH{=}N{-}NH_2 \xrightarrow{[O]} CH_3{-}O{-}C_6H_4{-}CH{=}\overset{+}{N}{=}\overset{-}{N}$$

p-methoxybenzylhydrazone *p*-methoxyphenyldiazoalkane

(4.9)

Method

Hydrazine sulphate (20 g) was dissolved in 400 ml of water at about 40 °C. *p*-Methoxybenzaldehyde (39 g) was added slowly, with stirring. The yellow floccular precipitate of the azine was collected by filtration, washed with ethanol, and dried *in vacuo*.

The azine (5.4 g) was dissolved in anhydrous hydrazine (4 g) and refluxed behind a protective screen for 16 h. Precautions were taken to prevent the admission of moisture. The pale-yellow reaction mixture was then poured into 50 ml ether. (The ether had previously been freed from acid by standing over pellets of NaOH and from peroxides by passage through an alumina column.) The ethereal layer was washed twice with water, dried over anhydrous Na_2SO_4, and stored, if necessary, in the absence of air.

The ethereal solution of the hydrazone was filtered into a conical flask in an ice bath and a freshly prepared saturated solution of KOH in ethanol (0.5 ml) was added. The solution was stirred and yellow HgO (17 g) was added in portions over 1.5 h. The solid became green and the solution red. Stirring was continued for 45 min after the last addition of HgO and the solution was then filtered. The filtrate (an approximately 0.2 M solution of the diazoalkane) was stored in a number of small vials at −20 °C. It decomposed under these conditions with a half-time of 2–4 weeks or more.

Notes

(1) The hazards of this preparation are probably similar to those of Example (4.4).

(2) The hydrazone is readily reoxidized to the azine and it is this reaction that gives the yellow tinge that the ethereal solution of the azine acquires on handling, or on storing in the presence of air.

(3) The diazoalkane can be prepared from the aldehyde by an appropriate modification of the method of Example (4.7).

(4) All the diazoalkanes (with the exception of diphenyldiazomethane, which was a crystalline compound and could be weighed) were assayed by reacting a measured sample with an ethereal solution containing an appropriate excess of benzoic acid or 3,5-dinitrobenzoic acid. The amount of acid esterified was determined either by isolating and weighing the ester or by back titration of unconsumed acid.

Approximate molar extinction coefficients at 500 nm were determined by drawing spectra of parallel samples to those assayed as above. The values for the coefficients were approximately 2 for the aliphatic diazoalkanes, approximately 24 for phenyldiazomethane, and about 40 for the *p*-methoxyphenyldiazoalkane. These values were found to be sufficient for rough working estimates of the concentration of reagent.

Example (4.7). The preparation and use of a water-soluble diazoalkane

Source. Offord *et al.* (1976).

Process involved.

$$CH_3{-}O{-}C_6H_3(SO_3Na){-}CHO \xrightarrow[\text{(anhydrous)}]{N_2H_4} CH_3{-}O{-}C_6H_3(SO_3Na){-}CH{=}N{-}NH_2 \xrightarrow{[O]}$$

aldehyde — hydrazone

$$CH_3{-}O{-}C_6H_3(SO_3Na){-}CH{=}\overset{+}{N}{=}\bar{N} \qquad (4.10)$$

diazoalkane

Method

Solid 4-methoxy-3-sulphobenzaldehyde (2.38 g) was crushed and placed in a beaker. To this was added 1.5 ml anhydrous hydrazine followed by 15 ml ethanol with very vigorous stirring. The liquid immediately took on an opaque white appearance. Stirring was continued until the initially viscous mixture was flowing freely. The mixture was then allowed to stand and the white precipitate (the sodium salt of the hydrazone) filtered off, dried *in vacuo*, and stored at −20 °C.

The 4-methoxy-3-sulphobenzaldehyde hydrazone (sodium salt) (0.255 g) was dissolved in 12.4 ml of dimethylformamide and 3.4 ml tetramethylguanidine added to the solution. The mixture was cooled in a bath of isopropanol/solid CO_2 and stirred magnetically. Lead tetraacetate (0.63 g), previously dried *in vacuo*, was added and the stirring continued for 10 min. Aqueous KOH (5%, w/v) was then added (3 ml) and, after stirring for 10 min, the whole mixture was filtered through Celite in a Buchner funnel (which had been pre-cooled by standing at −20 °C). The filtrate was collected in a pre-cooled Buchner flask. The clear, pink solution was stored at the temperature of isopropanol/solid CO_2. It had completely decomposed after 36 h.

The solution, prepared as above, was added dropwise to a vigorously stirred solution of benzoylarginine (54 mM in ethanol/water (17 : 3, v/v)) which was held between pH 5 and 5.5 with 0.5 M HCl in an autotitrator cell at 0 °C. After a 10-fold excess of reagent had been added (based on the hydrazone), the mixture was dried down and subjected to preparative electrophoresis at pH 3.5 (28 mg (including salts) in 10 μl per cm of Whatman 3 MM paper) to remove unreacted benzoylarginine. The ester was detected by staining a guide strip for arginine-positive material. It moved as a result of endosmosis only, while the benzoylarginine travelled toward the negative electrode. A test spot of the ethyl ester moved even more rapidly to the negative. The ester was eluted from the paper with pyridine/water (1 : 9, v/v).

Notes

(1) Offord *et al.* (1976) were unable to crystallize any of the products or intermediates, and relied for the characterization of the diazoalkane on its

spectroscopic properties, on its loss of colour with acid with evolution of approximately the expected volume of what appears to be N_2 and on the properties of the product formed by reacting it with benzoylarginine. These data are sufficient to establish that the product has the characteristics that are required of it to be useful in semisynthesis, but do not permit more than inferences —albeit rather obvious ones—as to its chemical structure. This Example is much more preliminary than the others in this book, but is offered in the hope that it will stimulate further, or parallel development of what may well turn out to be a very useful group of compounds.

(2) This method was developed by adapting a general low-temperature method for diazoalkanes proposed by Holton (1971). Attempts to prepare the reagent by the method of Example (4.6) were unsuccessful.

(3) The precursor aldehyde is commercially available.

(4) The lead tetraacetate was assayed by dissolving 0.1 g of the material as obtained in 15 ml glacial acetic acid. To this was added 0.5 g KI and 13 g sodium acetate in 65 ml water. (0.63 g of the tetraacetate = 0.0013 equivalents.) The iodine produced was titrated against 0.01 M $Na_2S_2O_3$ (2 equivalents of thiosulphate are required per equivalent of lead tetraacetate).

(5) The diazoalkane was assayed by manometric estimation (after absorption of CO_2) of the N_2 evolved on acidification. The molar extinction coefficient at 500 nm was, very roughly, 40.

(6) It would be very desirable to remove the excess base, and salts, from the solution of the diazoalkane before using it.

Example (4.8). Selective deprotection of α-COOH esters of tryptic peptides by trypsin

Source. Offord (1969a), Rees and Offord (1976a), and Offord *et al.* (1976).

Process involved. Specific catalysis of the hydrolysis of α-COOH esters of the carboxyl-terminal arginine of tryptic peptides that result from limited tryptic digestion.

Method

Protected peptide was dissolved in dimethylformamide/water (1 : 1, v/v) to a concentration of up to 10 mM and titrated to an apparent pH of 7 (glass electrode) with 0.05 M NaOH. The base was added slowly, with vigorous stirring, in order to avoid high local concentrations of NaOH. A solution of trypsin was made up in the same solvent to a concentration of 5 mg/ml. A 1 M solution $CaCl_2.6H_2O$ was added to the trypsin solution to a final concentration of 0.05 M and the pH of the solution was adjusted to 7. The trypsin solution (an enzyme–substrate ratio of 1 : 2500 on a molar basis—2 μl per μmol of peptide—was usually sufficient) was added to the peptide solution. The ensuing reaction was followed by observing the uptake of alkali in the autotitrator. Uptake had usually ceased (or fallen to background value due to CO_2 uptake) at 95–105% of the theoretical value for the liberation of one —COOH group. The reaction took from a few minutes to 2 h, depending on the peptide. The peptide was freed from trypsin and salts by gel filtration on lipophilic Sephadex LH-20 (cf. Example (3.8)) in dimethylformamide, ethanol, ethanol/water (17 : 3, v/v), or dimethylformamide/water (17 : 3, v/v) depending on the solubility of the peptide derivative. A column 2.2 cm × 105 cm was usually satisfactory for 50 mg of derivative.

The extent of specific deprotection can be checked by observing the electrophoretic mobility (Example (A.6)) of those derivatives that would still run on paper and the ability of carboxypeptidase B to liberate arginine from the carboxyl terminus (digestion was in 50 μl of 0.2 M *N*-ethylmorpholine acetate buffer, pH 8, 20–30% in pyridine, for 2 h at 37 °C with an enzyme–substrate ratio of 1 : 100, w/w).

Notes

(1) Rees and Offord (1976a) found that the water concentration below which esterolysis would not proceed varied from peptide to peptide, and from organic solvent. Even with Ca^{2+} present, trypsin tolerated the lower concentrations of water (70–50% and below) less well if the organic component was pyridine than if it was dimethylformamide. A few trials with dimethylsulphoxide suggested it would be at least as satisfactory an organic component as dimethylformamide. The report of Heinrich (1972) on the cleavage of model substrates suggests that *N,N*-dimethylpropionamide–water mixtures might be best of all.

(2) The pH is a little lower than the optimum value for trypsin in ordinary aqueous media. A pH of 7 was originally chosen to prevent the loss of side-chain amides belonging to very labile peptides (Offord and Rees, 1976a), but it is probably true that, in semi-aqueous solution, the optimum value is in any case close to 7.

(3) Certain susceptible peptides can be treated at an enzyme–substrate ratio of 1 : 5000 (molar basis). Such low ratios indicate how much more readily trypsin will promote esterolysis than it will proteolysis. Offord *et al.* (1976) have compared the rate of proteolysis of the bond Arg^{22}-Gly^{23} in *N*-protected insulin *p*-methoxybenzyl ester with that of α-esterolysis at the same position in *N*-protected des-(B23-B31) insulin *p*-methoxybenzyl ester (cf. the present Example with Example (3.5)). The more vigorous conditions required for proteolysis were sufficient to deprotect the α-COOH ester of alanine B30.

(4) See Note 4 to Example (3.5) for a possible alternative method for the trypsin.

(5) All the diazoalkanes described in this Chapter give esters that are susceptible to tryptic cleavage when on the α-COOH of arginine (and, presumably, lysine). This is surprising in view of the steric hindrance of the 1,1-dimethylethyl ester and the negative charge on the 4-methoxy-3-sulphobenzyl ester.

Example (4.9). The selective protection of the side-chain —COOH groups of an *N*-protected CNBr peptide

Source. Offord (1972), Wallace (1976), and Wallace and Offord (1979).

Process involved. Reactions (4.5) and (4.6). The terminal α-COOH group is protected from esterification as the lactone. It is liberated afterwards by base treatment.

Method

The peptide was taken to the lactone form and $—NH_2$-protected as described in Example (4.1). It was then esterified as described in Example (4.3). The peptide ester was then dissolved (if possible) in a 2% solution (v/v) of *N*-ethylmorpholine, taken to pH 9 with solid CO_2. A peptide concentration of 1 mM is suitable. The pH was then readjusted to 9, if necessary. Solubility could be enhanced by using a 1 : 1 mixture (v/v) of an organic solvent (dimethylformamide or dimethylsulphoxide) and water to make up the *N*-ethylmorpholine solution. (See also the Note on solubility, p. 218). The peptide was left in the basic solution for 1–2 h at 20 °C and freeze dried. The extent of opening of the lactone was checked by paper electrophoresis at pH 6.5 (Example (A.6)) for peptides that would run, and, for those that would not, by the extent of liberation of homoserine by carboxypeptidase digestion (see Example (5.3)).

Notes

(1) Offord (1972) used triethylamine-carbonate, pH 10, to open the lactone. Some esters are so base labile that they are partly removed by this treatment (Wallace and Offord, 1979). Susceptibility appears to depend on the sequence of neighbouring residues and exceptionally labile side-chain esters are partially cleaved even by the milder treatment given in the Method. Wallace and Offord (1979), using the *p*-methoxybenzyl ester and the methyl ester found that the unwanted products of this partial cleavage could be removed from the undamaged material by gel filtration on Sephadex LH-20—the separation was presumably based, not on molecular size, but on the discrimination often shown by gel-filtration media between molecules of differing degrees of hydrophobicity.

(2) Alternatively the peptide can be esterified in methanol/HCl (Example (2.10)) and then N^{α}-protected (Example (4.1)).

Example (4.10). The imidazole protection of an N^{α}-Boc- tryptic peptide

Source. Rees and Offord (1976a).

Process involved. Reaction (1.21) in which R = $(CH_3)_3C$—. The reagent, a trifluoroacetic acid ester, is first made from the parent alcohol.

Method

A solution of trifluoroacetic anhydride (0.35 ml) was made up in ice-cold, redist. pyridine (7.5 ml). This solution was added at 0 °C to 323 mg (1.5 mmol) 2,2,2-trifluoro-1-hydroxy-*N*-t-butyloxycarbonyl ethylamine. The solution, which turned pale brown during the addition, was allowed to stand at 25 °C for 30 min. Meanwhile, Boc-peptide (150 μmol) was freed of any possible N^{im}-Boc- groups by the method of Note 2 to Example (4.1). It was then added as a saturated solution in redist. pyridine and the mixture stirred for 24 h. If an electrophoretic check (Example (A.6)) showed that reaction was not complete, a further quantity of reagent (750 μmol) was made and allowed to react for another 24 h. When reaction was complete the mixture was taken to dryness on a rotary evaporator and extracted with ether to remove unwanted materials.

Notes

(1) This method is also applicable, in general outline, to proteins as well as peptides.

(2) The 1-hydroxy compound is commercially available. Rees and Offord (1976a) found its preparation by the method of Weygand *et al.* (1967) difficult to accomplish.

(3) Gel filtration of the reaction product would be preferable to extraction if it was desired to remove the very last traces of by-products.

CHAPTER 5

Modification of fragments

Once we have acquired a set of fragments in the required state of protection, the time has almost come to begin coupling them together. We must, though, pause to recall that the object of all our work was not to recreate the original molecule, but rather to produce a modified version of it.

Approaches to modification

One method of introducing a change would be to discard one of the natural fragments and replace it by one synthesized by conventional means, in the proper state of protection and with the modification included. Another approach would be to try to modify the fragment that contains the residue to be altered. This could take the form of exposing the fragment to one of the side-chain-directed groups that are in general use in protein chemistry for modification of the whole molecule. Such reagents have been well reviewed in their application to intact proteins—for example, by Means and Feeney (1971), by Glazer *et al.* (1976), and by Cohen (1968). Alternatively, the modification of the fragment could be carried out semisynthetically.

The first option, in which the fragment is discarded and a replacement synthesized, is frequently used but except in the simplest cases will not recommend itself to laboratories other than those that are well equipped for conventional synthesis. For fragments of any size, most other workers would at least wish to consider as an alternative the modification of the natural fragment.

Modification by means of side-chain reactivity

Many of the procedures for the chemical modification of side-chains can be applied to the chosen fragment more advantageously than to the whole protein. Only a few such reagents can be directed specifically and quantitatively to the side-chain of one particular residue in a whole protein. There are likely to be many other side-chains of the same sort and, unless we are dealing with the catalytic residue of an enzyme, only fortuitously will the chosen one have a usefully greater reactivity than the others. (Attempts to modify particular cysteine residues are least subject to problems as cysteine occurs rather sparsely in proteins and has a fairly distinctive type of reactivity.) But for most purposes, success is more likely if a fragment is used that contains only one or a very few of the reactive side-chain.

Example (5.1) describes a modification that depends on specific side-chain reactivity.

Side-chain modifications, whether carried out on a protein or a fragment, more often involve the *addition* of a chemical group to the existing bulk of the amino-acid residue. The groups that one might wish to add for eventual studies by fluorescence spectroscopy, electron-spin resonance, X-ray crystallography, or nuclear magnetic resonance (see below for examples) are often large. The addition even of only one such group can easily lead to a significant increase in dimensions of the protein. The group could easily be 10 Å across while the protein—usually a very compact structure, for all its molecular weight—could be as little as 30 Å in one or more of its dimensions. Even more seriously, the added groups are also likely to be apolar. In order to have reacted at all the side-chain to which the group is attached must have been polar; its masking could lead to a local, or even a general, denaturation.

Modification by semisynthesis

One must not undervalue such approaches as those that we have just discussed. Much of what we know about the relationship between structure and function in proteins was obtained by their use. But a semisynthetic modification would give us, in favoured instances, the opportunity to exchange any of the normal 20 amino acids for any other—a thing not possible if we rely on side-chain reactivities alone. Further, if we need to introduce a bulky group for any of the spectroscopic or other purposes mentioned above, we can synthesize an α-amino acid that has the group *as its side-chain*. The growth in bulk of the modified protein is then partially offset by the fact that the new group replaces a previously existing side-chain and is not added to it (see Fig. 5.1). If we are able to choose a residue with a large, apolar side-chain as the site of the modification—and as we do not rely on the reactivity of the original side-chain, as is often quite possible—then in favourable cases the damage to the balance of forces that stabilizes the protein's conformation may be negligible.

If, in comparison with the use of side-chain reactivity, we now have a wide range of sites at which it is technically possible to introduce a change, how are we to choose between them? If we need to change, say, a residue at the active site of an enzyme, the matter is more or less decided for us, but many other types of modification would be just as useful in one of a number of positions. Two types of data are particularly useful in helping us to decide where our alteration can be made so as to incur the minimum risk of disrupting the structure or function of the protein. First, if an X-ray crystallographic model is available, we can deduce what contacts would be made between the altered side-chain and neighbouring atoms and we can try to guess their effect. Second, if amino-acid sequences are available of homologues of our protein from other species, we can gain some clue as to the constraints upon the variation of each residue.

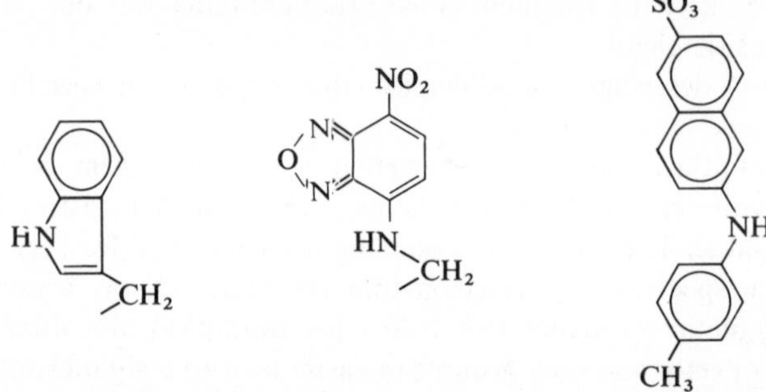

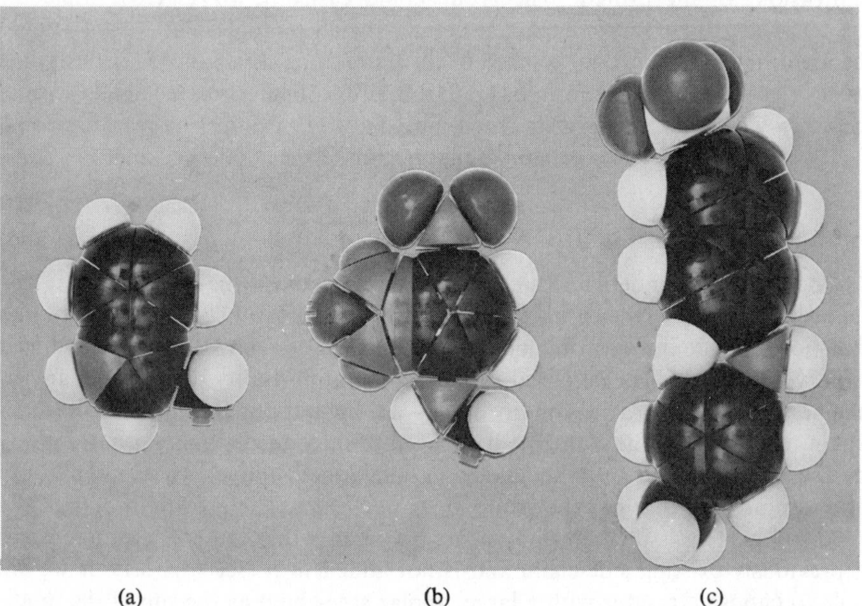

(a) (b) (c)

Fig. 5.1. Changes in bulk during protein labelling illustrated by space-filling models and structural formulae. (a) A tryptophan side-chain; (b) the side-chain of a synthetic, fluorescent amino acid which could replace it in the sequence; (c) 2-*p*-toluidinyl-6-naphthalenesulphonate, a fluorescent molecule which is commonly *added* to the structure of the protein, rather than being used to replace a side-chain.

Some sites are invariant across all species tested. Others only change in ways that conserve the functional abilities of the side-chain: a lysine might be replaced by an arginine, or a valine by an isoleucine. Of course, we cannot know if this apparent conservatism reflects some vital role for the residue or if it is merely the outcome of our having examined an insufficient number of species, but we would be well advised, if we have the freedom to choose, only to make functionally conservative alterations at such sites. We are on surer ground if we do find a site at which there has been a wide variation of character—exchanges between tryptophan, serine, and glutamic acid, say—and we should try if we can to reserve our more drastic changes for such positions.

We have to make use of these principles when deciding what to do about the fact that CNBr converts the side-chain of methionine (—$CH_2CH_2SCH_3$) to that of homoserine (—CH_2CH_2OH). For instance, X-ray studies and sequence comparison join with other evidence in making it quite clear that methionine-80 of horse-heart cytochrome *c* has a vital role in biological activity which homoserine could not possibly fulfil. We would therefore avoid all but the most subtle changes at this site—indeed we would almost always try to restore the proper side-chain at this position. (We shall see in a moment how this is done.) On the other hand, methionine-65 of the same protein is known to be replaced in one species by serine, a residue quite similar in its properties to homoserine; as the X-ray data would predict and as actual experiment confirms (Corradin and Harbury, 1974a), the substitution for homoserine at this point leaves the protein with biological activity.

Semisynthetic modification in practice

The scheme for modifying the fragment is much the same as the overall plan for modifying the protein. Since our fragments are already protected the first operation is cleavage, which is then followed by insertion of the synthetic portion. Semisynthetic modifications to fragments seem so far all to have been of the stepwise type. There is nothing to prevent our using a second exolytic agent on the fragment and thereby gaining access to more than the 33% of the residues mentioned on p. 55. No doubt examples of this approach will eventually be reported, but if the scheme were very elaborate it would begin to approach a conventional synthesis in its difficulty.

Cleavage

Example (5.2) gives practical details of the application of two cycles of the Edman reaction to a fragment. It resembles in principle Examples (3.1) and (3.2).

The details of Example (5.2) serve to remind us that almost all the operations described in this book take place under conditions that can promote the conversion of amino-terminal glutamic acid, or glutamine, to pyrrolidone carboxylic acid.

$$\begin{array}{c} \text{COOH} \\ | \\ \text{CH}_2 \\ | \\ \text{CH}_2 \\ | \\ \text{H}_2\text{N—CH—CO—} \end{array} \rightleftharpoons \begin{array}{c} \text{O} \\ \backslash\backslash \\ \text{C——CH}_2 \\ | \quad\quad | \\ \text{HN} \quad \text{CH}_2 \\ \text{CH—CO—} \end{array} + \text{H}_2\text{O} \qquad (5.1)$$

This might pass unnoticed (particularly if it occurs during α-NH_2 protection) or in the first stage of a cycle of the Edman reaction and if it goes toward completion it can frustrate the reaction scheme entirely. Some glutamyl- and glutaminyl-peptides are much more prone to this conversion than are others. If the problem is encountered, it is best to remove the residue at once by the Edman reaction, using the special precautions described in Note (3) of the Example, and subsequently replace it with one in a suitable state of protection. An alternative, which is only available if the residue at risk is glutamic acid, is to methylate (by the method of Example (2.10), not via diazomethane). Wallace and Offord (1979) found that little cyclization occurred either during methylation or on demethylation of the amino-terminal glutamic acid residue of a susceptible peptide.

There seem to be no reports of attempts to solve this problem by using an enzyme that specifically removes pyrrolidone carboxylic acid residues from the amino terminus of peptides and proteins.

Example (5.3) give details for the removal by carboxypeptidase A of the homoserine residue from the carboxyl terminus of CNBr fragment, a necessary operation when the substitution of methionine by homoserine cannot be tolerated (p. 93). The Example concludes with a description of the further truncation of the fragment by the use of carboxypeptidase B for the controlled removal of the new carboxyl-terminal residue, N^{ϵ}-acetimidyl lysine, which is susceptible to its action. Although there are several operations in the Example they are rapid and reasonably efficient; the overall effort involved in following this route compares very well with that required for the corresponding conventional synthesis.

Example (5.4) describes the use of carboxypeptidases for the modification of a protease inhibitor.

Active esters

The rest of the Chapter is devoted to the preparation of active esters (p. 8) of protected amino acids. Some of the amino acids involved are the ones found in natural proteins while others are of less familiar structure. The subsequent use of the active esters, in the reconstruction of the fragments, is considered in detail in Chapter 6.

Ordinary, protein amino acids

The methods for the preparation of the active esters of the ordinary amino acids are well known. The literature cited in the introduction to the Bibliography gives access to published accounts of the preparation of nearly every imaginable active ester. Example (5.5) describes the preparation of Boc-methionine hydroxysuccinimido ester. This is a typical example of the procedure, and the derivative itself is likely to be commonly used in semisynthetic work since it is often needed for the replacement of methionine when using CNBr fragments.

The insertion of a cysteine in a sequence has particular attractions. Example (5.6) describes the preparation of an active ester of a derivative with suitable α- and side-chain protection. Some proteins lack cysteine altogether or at least have none on their surface. It would be worthwhile to be able to insert a cysteine side-chain in a single, selected site on the surface of such proteins. It is not at present quite clear if conventional —SH-protecting groups are totally satisfactory for these purposes, since their methods of deprotection often seem to pose a threat to the integrity of, in particular, any existing disulphide bridges. Success would be well rewarded, since, because of the unique reactivity of the —SH group, specific substitutions could then be made after reassembly of the molecule. A substitution made in this way might be tolerated by the protein even if the change would have prevented the assumption of the correct tertiary structure if introduced during the synthesis. Modification after crystallization might lead to success in instances where modification at an earlier stage failed.

There are a large number of reagents that tend to be specific for the —SH group in proteins, and which are useful in introducing modifying groups of spectroscopic or crystallographic interest. Such reagents could be employed with semisynthetically introduced —SH groups as well as with natural ones.

Isotopically labelled amino acids

The need for isotopically labelled proteins is most often met by side-chain modification of the intact, native molecule. Iodination by ^{131}I or ^{125}I is particularly favoured as a means of obtaining radioactive derivatives and there are a variety of procedures (see, for example, Hunter, 1973) for carrying out the modifications. Although often easy to perform, the reactions tend to produce heterogeneous and sometimes partially inactivated products. Some, though otherwise mild in their effects, involve the loss of a functional group (e.g. Bolton and Hunter, 1973). All suffer from the disadvantage that they add a sizeable atom to the bulk of the protein. The iodine atom has approximately the bulk and apolarity of a benzene ring. The potential disadvantages are even more severe when the iodine is introduced as part of a large organic substituent group.

The semisynthetic insertion of an isotopically labelled amino acid can be quite a simple process and promises a product of authentic molecular structure. It has been used for the introduction of ^{3}H into insulin for tracer work (Halban and Offord, 1975) and stable isotopes, such as ^{13}C, for nuclear magnetic resonance studies (Saunders and Offord, 1972). The tritiation of insulin was a particularly apt case for semisynthesis—most ordinary methods of tritiation are reductive, and the disulphide bridges would be broken if they were used; their re-formation by oxidation is a notoriously inefficient process (see p. 161).

Conventional peptide syntheses normally employ active esters in quantities of the order of 100 mg–1 g or even more, while a reasonable quantity of radioactively labelled amino acid might well be measured in microgrammes. Example (5.7) explains how the difficulties presented by working at this very low level can be overcome.

Synthetic amino acids

Amino acids of unusual structure can be synthesized *de novo* by one of the usual routes (Greenstein and Winitz, 1961) as can isotopically labelled amino acids that are otherwise unobtainable. Example (5.8) illustrates such a synthesis, of *m*-iodophenylalanine.

Synthetic amino acids are obtained as racemic mixtures and must be resolved. Enzymic methods (Greenstein, 1957) are best and constitute an important series of ancillary techniques for those engaged in semisynthesis. Proteases can be used to catalyse the formation of an insoluble derivative. For example, papain assists in the precipitation of the phenylhydrazine from solutions of —NH_2-protected amino acids and phenylhydrazine. The enzyme shows a strong, but not absolute, preference for catalysing the reaction with the L-form.

$$\text{R–NH–CH(R}_1\text{)–COOH} + \text{H}_2\text{N–NH–C}_6\text{H}_5 \rightleftharpoons \text{R–NH–CH(R}_1\text{)–CO–NH–NH–C}_6\text{H}_5 + \text{H}_2\text{O} \quad (5.2)$$

It is sensible to use the —NH_2 protection that will be required in the subsequent semisynthesis.

The phenylhydrazide can be decomposed by controlled oxidation to give the free carboxyl compound.

$$\underset{\text{phenylhydrazide}}{\text{R–NH–CH(R}_1\text{)–CO–NH–NH–C}_6\text{H}_5} \xrightarrow{[O]} \underset{\text{phenyldiimide}}{\left[\text{R–NH–CH(R}_1\text{)–CO–N=N–C}_6\text{H}_5\right]}$$

phenyldiimide → $\underset{\text{deprotected derivative (major product)}}{\text{R–NH–CH(R}_1\text{)–COOH}} + N_2 + C_6H_6$

phenyldiimide → (side-reaction with another molecule of phenylhydrazide) → $(\text{R–NH–CH(R}_1\text{)–CO})_2\text{N–NH–C}_6\text{H}_5$

di-aminoacyl-phenylhydrazide (minor product) (5.3)

Example (5.9) illustrates the use of these two reactions with synthetic, DL-samples of the fluorophenylalanines—amino acids which are of use in nuclear magnetic resonance studies.

As an alternative to papain, one can use carboxypeptidase A to catalyse the freeing of the α-NH_2 group of a protected amino acid. The specificity of the enzyme requires that the α-COOH group must be free already.

$$\text{R}-\text{NH}-\underset{}{\overset{\text{R}_1}{\overset{|}{\text{CH}}}}-\text{COOH} \longrightarrow \text{H}_2\text{N}-\overset{\text{R}_1}{\overset{|}{\text{CH}}}-\text{COOH} \qquad (5.4)$$

Once again the enzyme shows a strong preference for the L-form, which is consequently liberated as the free amino acid. The unchanged derivative of the D-form can be separated from the free amino acid by precipitation or extraction. Example (5.10) gives experimental conditions for the use of carboxypeptidase for the resolution of DL-[^{13}C]phenylalanine, another amino acid of value in n.m.r. studies. The method has the disadvantage that the α-NH_2 of the L-form must be reprotected, but it is offset by the lack of a need for a process such as Reaction (5.3) and, in practice, the greater ease of obtaining the pure D-form. This derivative can be of value since the protein may well tolerate the introduction of a D-amino acid at the ends of the chain, and experiments with the derivative so formed may provide information which supplements that obtained with work on the derivative containing the L-form.

Semisynthetic amino acids

This approach to the preparation of amino acids of novel structure is not the fanatical over-insistence on the semisynthetic principle that it might seem. If a new side-chain is required that can be obtained by an easy chemical transformation of the functional group of a common L-amino acid, the approach has everything to recommend it. Provided the route chosen does not affect the chiral centre, one is spared the necessity for applying Reactions (5.2) and (5.3), or (5.4). For instance, we can use the Hofmann degradation to prepare the L-form of N^{α}-Boc-2,3-diaminopropionic acid in a single step, starting with N^{α}-Boc-L-asparagine.

$$\text{Boc}-\text{NH}-\overset{\overset{\text{CO}-\text{NH}_2}{|}}{\overset{\text{CH}_2}{\overset{|}{\text{CH}}}}-\text{COOH} \xrightarrow{\text{Br}_2,\text{NaOH}} \text{Boc}-\text{NH}-\overset{\text{CH}_2\text{NH}_2}{\overset{|}{\text{CH}}}-\text{COOH} \qquad (5.5)$$

The effort involved contrasts favourably with that needed to synthesize the DL-amino acid with specific protection of the α-NH_2 and to resolve it. The derivative is a useful one because it makes possible the synthesis of an amino acid with a side-chain which has useful properties for fluorescence studies in proteins (Moore, 1978b) (Reaction (5.6)). The Boc- derivative can then be converted to the active ester in the usual way. Unlike most usefully fluorescent groups, the side-chain is not much larger than that of an amino acid (Fig. 5.1).

Example (5.11) gives the experimental details, starting from Boc-L-glutamine.

$$\underset{\text{Boc}-\text{NH}-\text{CH}-\text{COOH}}{\overset{\text{CH}_2\text{NH}_2}{|}} + \text{NO}_2,\ \text{N},\ \text{O},\ \text{N},\ \text{Cl} \longrightarrow \text{NO}_2,\ \text{N},\ \text{O},\ \text{N},\ \text{NH}-\text{CH}_2-\underset{\text{Boc}-\text{NH}-\text{CH}-\text{COOH}}{} \qquad (5.6)$$

Another useful starting compound is L-serine, as the side-chain —OH can be readily substituted by a halogen which can in its turn engage in many synthetic reactions. Many similar transformations of the amino acids are known (for a review, see Greenstein and Winitz, 1961) and others will suggest themselves.

We saw in Example (5.8) a rather troublesome synthesis of *m*-iodophenylalanine. The *p*-iodo derivative can be made more simply from L-phenylalanine (Krail *et al.*, 1973).

$$\text{R}-\text{NH}-\text{CH}(\text{CH}_2\text{C}_6\text{H}_5)-\text{COOH} \xrightarrow{\text{nitration}} \text{R}-\text{NH}-\text{CH}(\text{CH}_2\text{C}_6\text{H}_4\text{NO}_2)-\text{COOH} \xrightarrow{\text{reduction}}$$

$$\text{R}-\text{NH}-\text{CH}(\text{CH}_2\text{C}_6\text{H}_4\text{NH}_2)-\text{COOH} \xrightarrow{\text{HNO}_2,\text{KI}} \text{R}-\text{NH}-\text{CH}(\text{CH}_2\text{C}_6\text{H}_4\text{I})-\text{COOH} \qquad (5.7)$$

The final step has been adapted to the submicrogramme level (C. Herrera-Caravajal, thesis in preparation). This makes it possible, if one must use iodination rather than tritiation in protein work (e.g. because of the very high specific activities possible with ^{125}I and ^{131}I), to introduce one atom only, and that at a site chosen to affect the structure and function of the molecule to the minimum possible extent.

Reassembly

The active esters may now be used to replace the residues removed from the fragment. For instance, in Example (5.2), the CNBr fragment of N^{ϵ}-acetimidyl cytochrome *c* that corresponds to residues 81–104 was truncated by the removal of isoleucine-81 and phenylalanine-82. Residue Phe^{82} is of interest in studies on the relationship between structure and function in cytochrome *c* and it might be profitable to use the active esters of fluorophenylalanine, isoleucine, and methionine in a stepwise synthesis to produce a fragment that begins H_2N—Met-Ile-(F)Phe-. In practice the starting sequence received side-chain —NH_2 protection before cleavage from the protein and so the product is ready at once to be coupled with the CNBr fragment adjacent to it in the sequence (which corresponds to residues 66–80). That fragment should first have had its carboxyl-terminal homoserine removed (Example (5.3)). The net effect of the operations that we have described is to restore the methionine residue, destroyed by CNBr, to its proper place in a modified sequence with the minimum of effort.

Coupling is an art in itself, and a discussion of the practical and theoretical issues involved is best left until the next Chapter.

EXAMPLES

Example (5.1). The conversion of serine to cysteine in peptides and proteins

Source. Photaki and Bradakos (1965), Neet and Koshland (1966) and Polgar and Bender (1966).

Process involved.

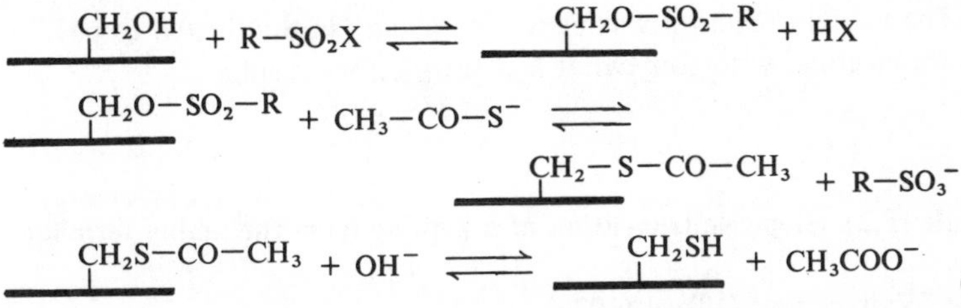

$$\underline{\overset{CH_2OH}{|}} + R{-}SO_2X \rightleftharpoons \underline{\overset{CH_2O{-}SO_2{-}R}{|}} + HX$$

$$\underline{\overset{CH_2O{-}SO_2{-}R}{|}} + CH_3{-}CO{-}S^- \rightleftharpoons \underline{\overset{CH_2{-}S{-}CO{-}CH_3}{|}} + R{-}SO_3^-$$

$$\underline{\overset{CH_2S{-}CO{-}CH_3}{|}} + OH^- \rightleftharpoons \underline{\overset{CH_2SH}{|}} + CH_3COO^- \qquad (5.8)$$

R—X = *p*-toluenesulphonyl chloride or phenylmethanesulphonyl fluoride. The final, hydrolytic step takes place spontaneously at neutral pH.

Method

Cbz-Gly-Ser-Gly-ethyl ester (0.38 g, 1 mmol) was dissolved in 3 ml of anhydrous pyridine, which had been pre-cooled to −5 °C. *p*-Toluenesulphonyl chloride (0.48 g) was added, and the mixture allowed to stand at between −5 °C and 0 °C for 1 h. Cold water (20 ml) was then added, with stirring. The crystalline *S*-tosyl peptide crystallized out on seeding and scratching.

The *S*-tosyl compound (0.135 g, 0.25 mmol) was dissolved in a solution made by adding 0.175 ml triethylamine to 0.1 ml thioacetic acid in 1 ml of ethyl acetate. After 24 h at 20 °C the solution was diluted with ethyl acetate.

This solution was washed with cold 0.5 M H_2SO_4, then with water, then with aqueous $KHCO_3$, and then with water again. The organic layer was dried over anhydrous Na_2SO_4 and evaporated *in vacuo*. The yield of crude product was 80% of the theoretical.

Notes

(1) Analytical data for the product are given in Photaki and Bradakos (1965). It can be purified by recrystallization from acetone/petroleum ether.

(2) Zioudrou *et al.* (1965) describe an analogous method.

(3) As given, the method is designed for totally synthetic peptides, but is easy to adapt for natural materials. The reaction scheme can be applied to proteins in aqueous solution (Neet and Koshland, 1966; Polgar and Bender, 1966). Both pairs of authors took advantage of the exceptional reactivity of the active-site serine of a protease to carry out the first stage of the reaction on the side-chain of just that residue without affecting any other serine residues in the protein. Phenylmethanesulphonyl fluoride was found to be faster for the first step than the tosyl chloride. Typical conditions were 1.2 mM phenylmethane-sulphonyl fluoride, 1 mM subtilisin at pH 7 in 0.1 M phosphate buffer at room temperature for 2 h. The solution was then made 0.7 M with potassium thio-acetate and adjusted to pH 5.5. The product was isolated by gel filtration after the mixture had been allowed to stand for 2 days at room temperature. The yield of cysteinyl product was about 80%. Less than 2–3% of the material reverted to the seryl form during the reaction.

(4) The protein reaction was also tried using Na_2S adjusted to pH 7 in place of the thioacetate, with somewhat less satisfactory results.

Example (5.2). Stepwise truncation of a peptide from the amino terminus

Source. Wallace and Offord (1979).

Process involved. Reactions (3.1) and (3.2) applied twice to Ile-Phe-Ala-Gly-Ile-Lys(Acim)-Lys(Acim)-Lys(Acim)-Thr-Glu-Arg-Glu-Asp-Leu-Ile-Ala-Tyr-Leu-Lys(Acim)-Lys(Acim)-Ala-Thr-Asn-Glu. This peptide is a CNBr fragment of cytochrome *c* and corresponds to residue 81–104 of the protein.

Method

Peptide (1.8 μmol, prepared as described in Example (3.7)) was dissolved in 4 ml pyridine/water (1 : 1, v/v). To this was added 2 ml of phenyl isothiocyanate (redist.)/pyridine (1 : 19, v/v, equivalent to a 400-fold molar excess over the α-NH_2) and the reaction mixture held at 37 °C for 3 h. The solution was dried *in vacuo* and treated with anhydrous trifluoroacetic acid (1 ml) for 30 min at

37 °C. The acid was then removed *in vacuo*. The dried material was suspended in 2 ml of water and extracted three times with 2 ml lots of n-butyl acetate. The aqueous layer was freeze dried. The whole cycle of reactions was repeated and checked by amino-acid analysis and end-group determination (Examples (A.1) and (A.2)).

Notes

(1) The necessary excess of the Edman reagent depends on the peptide and is best found by a trial on a small quantity of material. A 200-fold excess or even less is often sufficient but there seems to be no harm in using somewhat more than the minimum necessary amount.

(2) The n-butyl acetate removes the phenylthiohydantoins and, with the peptide used in the method above, leaves a clear aqueous solution.

(3) Peptides that have a tendency to undergo Reaction (5.1) should be handled as little as possible before beginning the Edman cleavage. For example, peaks 5 and 7 from the 7% formic acid column in Example (3.7), Method B, consist of peptides with an amino-terminal residue of glutamic acid which is very susceptible to the conversion and should not be further purified by re-application to the column until after the residue has been removed by the Method above. In very susceptible peptides conversion to the pyrrolidone derivative can take place even under the conditions of the phenylthiocarbamoylation, and it may be necessary to accelerate the desired reaction by increasing the concentration of the Edman reagent.

Should partial conversion to the pyrrolidone carboxylic acid occur, the only undesirable consequence is a loss of usable material. The final semisynthetic protein will not be contaminated by incorrect material. The conversion is, to all intents and purposes, an irreversible protection of the α-NH_2 group and will prevent any further reaction, including coupling, from occurring. Nonetheless, ion exchange could be used, if desired, to separate peptide that had undergone the conversion from material that had been successfully truncated, since the latter will have a free α-NH_2 group.

Example (5.3). The stepwise truncation of a CNBr peptide from the carboxyl terminus

Source. Wallace and Offord (1979).

Process involved. Removal of the carboxyl-terminal homoserine from Boc-Glu(OMe)-Tyr-Leu-Glu(OMe)-Asn-Pro-Lys(Acim)-Lys(Acim)-Tyr-Ile-Pro-Gly-Thr-Lys(Acim)-homoserine by carboxypeptidase A, followed by removal of the new carboxyl-terminal residue (N^{ϵ}-acetimidyl lysine) by carboxypeptidase B. The peptide represents residues 66–80 of horse-heart cytochrome *c*.

Method

The N^{ϵ}-acetimidyl peptide (peak 7, Example (3.7), Method B; 10 μmol) was esterified by the method of Note 3 to Example (2.10). The α-NH_2 group was then protected by the method of Example (4.1) and the lactone opened by dissolving the material in 0.02 M *N*-ethylmorpholine, adjusted to pH 9 with solid CO_2, for 2 h at 20 °C. The freeze-dried product was then dissolved in 0.2 M *N*-ethylmorpholine acetate, pH 7.5, and a solution of carboxypeptidase A was added (0.75 mg (37.5 i.u.) per μmol of peptide in the original sample, dissolved as described in Note 6). The reaction was allowed to proceed at 20 °C for 2 h 45 min and the mixture freeze dried. Gel filtration (Sephadex LH-20, 2.5 cm × 100 cm run in dimethylformamide/water (17 : 3, v/v) at 80 ml/h) was used to free the des-homoserine peptide from enzyme and peptide that had been partially demethylated in the lactone-opening step (both of which emerged in the void volume) and from homoserine (which emerged after the wanted product).

The peptide (approximately 6 μmol) was recovered from the column effluent by freeze drying. It was dissolved in 2 ml of 0.1 M NaCl/0.025 M Tris-HCl, pH 7.65, and carboxypeptidase B (0.43 mg (60 i.u.) in 0.21 ml 0.1 M NaCl) was added. The reaction was allowed to proceed at 20 °C for 40 min. The peptide solution was diluted with 12 ml dimethylformamide and purified by gel filtration as before. The freeze-dried product (approximately, 4 μmol) was checked by amino-acid analysis (Example (A.1)).

Notes

(1) The peptide, as it is obtained from the 7% formic acid column in Example (3.6), is about 25% in the open, as opposed to the lactone, form. It would be possible to convert it all to the lactone form before esterification—by dissolving in anhydrous trifluoroacetic acid (5 μmol/ml) for 1 h at 20 °C (see Note 1 to Example (4.1))—but in practice the acidic conditions of the esterification appear to achieve the conversion without any α-esterification. (The evidence for this statement is that the complete interconversion between the lactone and the open forms is still possible when the esterified peptide is examined and that the homoserine residue is quantitatively removed from it by the carboxypeptidase A.)

(2) The side-chain ester of Glu^{69} may be unusually labile to saponification. Certainly, about 25% of the peptide diester is converted to a form with the properties of the monoester even by the very mild treatment that opens the lactone (cf. Note 1 to Example (4.9)). This is not too serious as the monoester form can be removed on a gel-filtration column, probably as a result of hydrophobic interactions with the column matrix (Wallace and Offord, 1979). The saponification probably does not involve Reaction (1.20) (Wallace and Offord,

unpublished observations). The waste of material is regrettable, but the monoester could in theory be recycled.

(3) *N*-Boc-peptides cannot be run on the 7% formic acid column of Example (3.7) because the solvent causes a slight loss of protection.

(4) Should it be necessary to saponify the peptide before coupling with it as an amino component, the conversion to the pyrrolidone carboxylic acid must be prevented by keeping on the α-NH_2 protection until afterwards.

(5) The conditions for the α-NH_2 protection do not open the lactone or remove —CH_3 groups.

(6) The required quantity of the stock solution of carboxypeptidase A was centrifuged and the pellet was washed once with water. The pellet was dissolved in 2 M NH_4HCO_3 (to the same volume as the original suspension) and diluted by the addition of a further 9 vol. of water. The enzyme solution was added directly to that of the peptide.

(7) The carboxypeptidases perform less well in semi-aqueous solvents than does trypsin. They are particularly adversely affected by pyridine.

(8) The times of digestion with the carboxypeptidases, and the quantities required, are strongly dependent on sequence, and must be established by trial experiments. Electrophoresis at pH 1.9 and quantitative amino-acid analysis (Example A.1)) are extremely useful in assessing the extent of digestion. In the present instance, if economy is important, quite satisfactory results can be obtained with smaller quantities of the enzymes (one-half the suggested quantity of carboxypeptidase A in a 3-h digest; one-third the suggested quantity of carboxypeptidase B in a 2-h digest).

The sequence of the peptide in this Example is particularly favourable to the controlled removal of the two residues. If Lys^{79} were unprotected lysine, carboxypeptidase A would probably remove it and, at least, Thr^{78} as well as the homoserine derived from Met^{80}. N^{ϵ}-Acetimidyl lysine somewhat resembles arginine in structure and, as we hoped, in its resistance to carboxypeptidase A.

There will be very many sequences for which it will be possible to remove just the carboxyl-terminal homoserine, even if controlled degradation beyond that point could not be achieved. The digest would have to be stopped before other residues begin to be cleaved. If the succeeding residue or residues are particularly susceptible to the enzyme (see Ambler, 1967) it may be necessary to interrupt the digestion before the homoserine has been removed from all the molecules in the sample. If this is the case, the uncleaved material could probably be separated from the cleaved material by making use of the ability of the former to be converted to the lactone. The lactone might then be differentiated by means of its chemical or ion-exchange properties.

(9) It is possible to methylate the cytochrome *c* before CNBr cleavage. If such material is employed, the methylation given in the method above is unnecessary. Wallace (1976) used this approach to prepare des-homoserine forms of the peptide used above and of the much longer fragment corresponding to residues 1–80 of the protein (peak 3, Example (3.7), Method B).

Example (5.4). The use of carboxypeptidase B to replace an arginyl by a lysyl residue in a protease inhibitor

Source. Sealock and Laskowski (1969).

Process involved. The Arg^{64}–Ile^{65} bond of soyabean trypsin inhibitor was cleaved by trypsin. Arg^{64} was then removed by carboxypeptidase B. Its replacement with lysine was then catalysed by carboxypeptidase B in the presence of free trypsin.The reaction was pulled over in the direction of synthesis by the formation of a tightly bound complex between the product and the trypsin.

Method

The des-Arg^{64} inhibitor was prepared as described in the original paper. It was dissolved in 0.15 M KCl/0.03 M $CaCl_2$ to a concentration of 100 μmol/l. Trypsin (200 μmol/l) and lysine (70 mmol/l) were also present. The pH was adjusted to pH 6.7, carboxypeptidase B added to a concentration of 100 μmol/l, and the pH readjusted if necessary to 6.7 and held at that value subsequently. All pH adjustments were effected by the addition of dil. HCl or dil. KOH. After 10 days at 21 ± 1 °C the reaction was terminated by gel filtration on Sephadex G-75 (2.5 cm × 90 cm, for a loading of 90 mg of inhibitor) in 0.01 M Tris/0.1 M NaCl, pH 8.5. The first peak to emerge, which contained the complex between trypsin and the inhibitor, was dialysed and freeze dried.

Notes

(1) The above conditions were found by a series of trial experiments, monitored by gel electrophoresis (the product has one more positive charge than the starting material). It was found that the high concentration of carboxypeptidase B (it is considerably greater than the concentration required to remove Arg^{64} in the first place) was essential to success. The rate of the process was depressed by too high a concentration of free amino acid. If the free amino acid was left out altogether there was still some product formed, presumably from free amino acid liberated during autolysis of the trypsin.

(2) The authors report that the activity of the carboxypeptidase does not decrease significantly during the several days' incubation. It is remarkable that it should be able to withstand such prolonged exposure to trypsin. By contrast, the trypsin itself undergoes an appreciable degree of autolysis.

(3) The product, the trypsin-inhibitor complex, can be dissociated (with simultaneous re-formation of the Lys^{64}–Ile^{65} bond, see p. 189) by 6 M guanidinium hydrochloride followed by gel filtration in the same solvent (A. Morawiecki, quoted by Sealock and Laskowski, 1969).

(4) The transformation of the trypsin inhibitor to a chymotrypsin inhibitor has been effected through catalysis by carboxypeptidase A of the substitution of tryptophan for Arg^{64}. The reaction was pulled over to completion by having

chymotrypsin present, rather than trypsin (Leary and Laskowski, 1973). The full experimental details have not yet been published.

Example (5.5). The preparation of —NH_2-protected active esters of methionine

Source. Dekker and Fruton (1948), Schwyzer *et al.* (1959), and Anderson *et al.* (1964). Details from Saunders (1974) and Saunders and Offord (1977a).

Process involved. Reaction (1.12) in which R = $(CH_3)_3C$— or $C_6H_5CH_2$—, followed by Reaction (1.10) in which R = (succinimide ring: O, N—, O)

Method

N^{α}-Cbz-L-methionine. L-Methionine (1.49 g, 10 mmol) was dissolved in 2 M NaOH (5 ml) and cooled in an ice bath. Benzyloxycarbonyl chloride (1.67 ml, 10 mmol) and 2 M NaOH (5 ml) were added in 5 portions over the course of 30 min at 0 °C. The reaction mixture was stirred continuously during this period and for a further 1 h at 20 °C. The solution was extracted with ethyl acetate to remove unreacted reagent and then brought to pH 3 with conc. HCl. The product separated as an oil, yield 40% (Cbz-DL-methionine crystallizes).

Boc-methionine. L-Methionine (264 mg, 1.77 mmol) was stirred with MgO (225 mg) in a mixture of water (5 ml) and dimethylformamide (9 ml) at 40–50 °C until crystals of amino acid were no longer visible. To this suspension was added t-butyloxycarbonyl azide (0.6 ml) and the reaction allowed to proceed for 3 days at 20 °C. Very little ninhydrin-positive material remained. The MgO was spun down and washed with water. The combined supernatants were dried *in vacuo* and the residue dissolved in water. The aqueous solution was adjusted to pH 2.5–3 by the careful addition of 6 M HCl and extracted three times with ethyl acetate (20 ml). The combined organic layers were dried over anhydrous Na_2SO_4 and the solvent removed *in vacuo*.

Esterification. The —NH_2-protected methionine (1 mmol) and *N*-hydroxy-succinimide (150 mg, 1.3 mmol) were dissolved in 4.5 ml ethyl acetate. Dicyclo-hexylcarbodiimide (227 mg, 1.1 mmol) was dissolved in 5.5 ml ethyl acetate. Both solutions were cooled to 4 °C, mixed, and stirred at 4 °C for 18 h. A white precipitate of dicyclohexylurea was formed during the reaction, and was removed by centrifugation. The precipitate was washed with a small volume of ethyl acetate. The combined supernatants were extracted with 2 vol. of $NaHCO_3$ (5%, w/v), then 2 vol. of citric acid solution (1%, w/v), and finally with 2 vol. of water. The organic layer was dried over anhydrous Na_2SO_4 and the solvent removed *in vacuo*. The product was recrystallized from isopropanol, m.p. 129–133 °C (lit. 128 °C). Yield about 80%. The progress of the reaction and the quality of the product were checked by the thin-layer system of Example (A.9).

Notes

(1) This method can be used for the preparation of active esters of many other amino acids. The —NH_2-protected compounds can be purchased, or prepared by standard methods, e.g. that of Schnabel (1967).

(2) Some amino-acid derivatives are insufficiently soluble in ethyl acetate. Dimethylformamide or redist. 1,2-dimethoxyethane can be used instead. In such cases the solvent is removed *in vacuo* at the end of the esterification and replaced by ethyl acetate for the aqueous extractions.

(3) Isopropanol is the solvent of choice for recrystallization, but other solvents are sometimes required (e.g. chloroform/petroleum ether, 40–60 °C, for the hydroxysuccinimide ester of Boc-valine). The original literature should be consulted in cases of doubt.

(4) The ethyloxycarbonyl derivative for use in Examples (2.4) and (2.8) can be prepared by the Method above. The product is not crystalline (Saunders and Offord, 1976a).

(5) The dicyclohexylcarbodiimide should have a m.p. of 34 °C. It can be purified by sublimation. Care is necessary to avoid contact between it and the skin. The *N*-hydroxysuccinimide can be recrystallized from ethyl acetate, m.p. 99–100 °C.

(6) Moroder *et al.* (1976) describe the use of di-t-butyloxydicarbonate with excellent results.

Example (5.6). The preparation of Boc-,*S*-acetamidomethyl cysteine trichlorophenyl ester

Source. Marbach and Rudinger (1974), and Veber *et al.* (1972).

Process involved.

$$\underset{H_2N-CH-COOH}{\overset{CH_2SH}{|}} + CH_3-CO-NH-CH_2OH \rightleftharpoons \underset{H_2N-CH-COOH}{\overset{CH_2-S-CH_2-NH-CO-CH_3}{|}} + H_2O \qquad (5.9)$$

Method

S-Acetamidomethyl cysteine. Cysteine hydrochloride (1.576 g, 10 mmol) was dissolved in 10 ml redist. trifluoroacetic acid together with 0.89 g *N*-hydroxymethylacetamide (10 mmol) and stirred for 30 min at 20 °C. The mixture was then rotary evaporated to dryness and the residue dissolved in 1 M HCl. The solution was dried down, more 1 M HCl added, and the solution dried down again. The product was recrystallized from isopropanol, the crystals filtered by suction and washed with diethyl ether. The yield of the hydrochloride was 1.62 g (71% of the theoretical), m.p. 155–157 °C. A second recrystallization, from water/isopropanol, raised the m.p. to 166–168 °C.

N^{α}-Boc-, S-acetamidomethyl cysteine. Some of the product of the previous step (0.684 g) was dissolved in redist. dimethylformamide (6 ml) and the solution gassed thoroughly with N_2. Then, still under N_2 and at a temperature of 25 °C, 0.69 g tetramethylguanidine was added dropwise over a period of 10 min, alternating with 0.47 g t-butyloxycarbonyl azide mixed with a further 0.345 g of the base. The mixture was stirred overnight at 20 °C. The reaction mixture was concentrated *in vacuo* at 30–35 °C and 2 ml water added. The aqueous solution was twice washed with 2 ml portions of ethyl acetate and, after cooling in an ice bath, washed with a 50% aqueous solution of citric acid, saturated with NaCl. The aqueous layer was extracted twice with 3 ml ethyl acetate. The combined ethyl acetate layers were washed twice with 3 ml lots of saturated NaCl, dried with Na_2SO_4, and evaporated *in vacuo*. The oil residue was taken up in 0.4 ml diethyl ether. Crystals formed on gently warming and seeding the solution. Yield approximately 0.4 g, m.p. 110–112 °C. The m.p. can be raised by recrystallization from ethyl acetate/benzene.

Formation of the active ester. N^{α}-Boc-,*S*-acetamidomethyl cysteine (73 mg, 0.25 mmol), 2,4,6-trichlorophenol (60 mg, 0.3 mmol), and dicyclohexylcarbodiimide (52 mg, 0.275 mmol) were added in turn to 1 ml of dichloromethane held at 0 °C. The solution was then allowed to warm up to 20 °C. Crystals of the dicyclohexylurea soon began to form, and the mixture was left at 20 °C for 16 h. A check by the thin-layer system of Example (A.9) showed that the N^{α}-Boc-,*S*-acetamidomethyl cysteine ($R_f \simeq 0.2$, streaked) had been converted to the active ester ($R_f = 0.5$). The reaction mixture was dried on the rotary evaporator, resuspended in approximately 2 ml ethyl acetate, and centrifuged. The pellet was washed in a further 2 ml of ethyl acetate. The combined supernatants were filtered through two thicknesses of Whatman No. 1 paper and dried on the rotary evaporator, the solid was redissolved in 1 ml ethyl acetate at 50 °C. Petroleum ether (6 ml of 60–80 °C) was slowly added. The mixture was rewarmed to 50 °C after each addition. The solution was allowed to cool to room temperature. Crystals soon began to appear. After 2–3 h at 20 °C, the vessel was placed at 4 °C overnight. The crystals were separated from the supernatant by suction and dried *in vacuo*. The m.p. was 134–135 °C (expected, 137–138 °C). Yield 81%.

Notes

(1) The dicyclohexylurea cannot be centrifuged down from the reaction mixture in dichloromethane as the solvent is too dense.

(2) No doubt alternative methods for the introduction of the Boc- group (e.g. those mentioned in the previous Example) would work satisfactorily. It would probably still be desirable to exclude oxygen during the reaction when operating on cysteine derivatives.

(3) Some workers prefer the trichlorophenyl ester, used here and in Example (5.10), to the *N*-hydroxysuccinimido ester.

Example (5.7). The synthesis, at 20 Ci/mmol of Boc-[^{3}H]phenylalanine hydroxysuccinimido ester

Source. Halban and Offord (1975).

Process involved. Reaction (1.12) in which R = $(CH_3)C$— and Reaction (1.10)

O

in which R = N—.

O

Both processes take place at the μg level.

Method

The stock solution of [^{3}H]phenylalanine (up to 5 ml, which may contain, depending on the specific radioactivity, as little as 25 nmol but more normally 250 nmol) was freeze dried in a small, round-bottomed flask. The dried material was carefully washed into a Pyrex tube, with the use of a total of 1 ml of distilled water. The tube was constructed by drawing the end of a heavy-wall test tube (internal dimensions: 90 mm long × 10 mm internal diameter) to a point. The taper was such that the full internal diameter was reduced to zero over a distance of 35 mm. The solution was once again freeze dried. The use of heavy-wall tubing makes it possible to spin the tube in a bench centrifuge if, after any of the drying stages, the resuspended material is found to be excessively contaminated with particulate matter. The radioactive material was brought to a concentration in water of 25 nmol/μl, cooled on ice, and 1.5 vol. each first of redist. dimethylformamide (p. 218) and then of ethanol were added. The final concentration of amino acid was 6.25 nmol/μl. Redist. *N*-ethylmorpholine was added through a microcapillary until the externally indicated pH (p. 217) of a 0.1 μl sample of the solution was approximately 8.5. The quantity of base required was about 1 μl/100 nmol of amino acid and little harm seems to be caused by moderate excess. t-Butyloxycarbonyl azide was then added (50 nmol/nmol of amino acid) and the reaction vessel was stoppered. The mixture was dried down after 20 h at 37 °C and 0.1 μl of 0.05 M HCl was added per nmol of amino acid. The N^α-Boc-phenylalanine was extracted with 3 × 4 vol. of ethyl acetate. The extracted material was compared with authentic, non-radioactive standards by thin-layer chromatography on the system of Example

(A.9). The radioactive material was located with a scanning, gas-flow detector equipped for windowless counting. Phenylalanine has $R_f = 0$ while the protected amino acid has $R_f = 0.1–0.2$ in the (—COO^-) form and 0.5–0.6 in the (—COOH) form. The material moved as a single spot $R_f = 0.5–0.6$ and yields of better than 90% were obtained.

The ethyl acetate layer was dried down exhaustively *in vacuo* to free it from all traces of volatile reagents. It was then dissolved (25 nmol/μl) in dry 1,2-dimethoxyethane which contained an equivalent quantity of *N*-hydroxysuccinimide. After cooling to 2–4 °C an equal volume of dry 1,2-dimethoxyethane, which contained 1.1 equivalents of dicyclohexylcarbodiimide, was added. After 20 h at 2–4 °C the esterification was checked by the thin-layer system and had gone to completion (a single radioactive spot $R_f = 0.7–0.8$). Crystals of dicyclohexylurea were visible in the mixture, which was used at once, without being dried down (Example (6.3)).

Notes

(1) The small scale of the reaction is dictated by the cost of the [^{3}H]amino acid. It was found essential to scale down the quantities of other reactants, although it seems at first sight that these could be used in larger, more convenient amounts.

(2) Every effort should be made to avoid losses of [^{3}H]amino acid during the transfer and resuspension steps. Those parts of the walls of glassware on which dried material has been deposited should be rinsed down with particular care, while those parts that were not in contact with the solution before drying down should be left alone.

(3) The acid used in the extraction of the N^α-Boc-phenylalanine is thought to be necessary to ensure that the derivative is in the carboxyl (—COOH) form to facilitate extraction into an organic solvent and for the subsequent activation and to increase the efficiency of partition.

(4) Small departures from the method, particularly in the —NH_2 protection, led to drastically reduced yields, as did the failure to remove any gross contamination with insoluble material at the beginning. As stated in Note 1, it is essential to scale down the quantities of all reactants, not, as might be thought, that of the amino acid alone.

(5) It is usually undesirable to concentrate or to dry down materials of high specific radioactivity. Although the half-life of the radioisotope is long, secondary, chemical decomposition is greatly accelerated by the energy released by the radioactive decay. In particular the absorption of the β particles in the nearby medium leaves a large number of highly reactive ions and free radicals. Many more molecules are destroyed by this means, particularly at high concentrations, than decompose as a result of the decay of an isotopic atom in their own molecular structure. The ^{3}H-labelled active ester decomposes if it is dried down. It is not known for certain if this is a radiochemical phenomenon, but it does

not occur when the method is applied to slightly larger quantities of non-radioactive material.

It would be preferable not to have to dry down the starting material so often. As the solutions in question are all aqueous, they can be freeze dried which lessens (but does not abolish) the radiochemical damage.

(6) In the event of partial esterification, the thin-layer system may well show the unesterified product in the $R_f = 0.1$–0.2 position, corresponding to the ($—COO^-$) form.

(7) As with preparations on a larger scale, di-t-butyloxydicarbonate can be used instead of the azide for the amino protection of [^{3}H]phenylalanine. The conditions are the same as those given above for the azide, except that 4 nmol of the dicarbonate were added per nmol of amino acid. The yield of protected product is usually better than with the azide after an incubation of 20 h at 37 °C. Trial experiments on similar quantities of non-radioactive phenylalanine (J. G. Davies, private communication) show that, as would be expected, larger excesses of the dicarbonate reduce the time necessary for the reaction to go to completion. By contrast, we find that preparations using the azide do not respond to alterations in conditions in such a simple manner.

Example (5.8). The preparation of DL-*m*-iodophenylalanine

Source. Borrás (1972) based on Counsell *et al.* (1968).

Process involved.

$$m\text{-}I\text{-}C_6H_4\text{-}CH_3 \xrightarrow{Br_2} m\text{-}I\text{-}C_6H_4\text{-}CH_2Br \xrightarrow{CH(CO\text{-}O\text{-}CH_2\text{-}CH_3)_2\text{-}NH\text{-}CO\text{-}CH_3}$$

$$m\text{-}I\text{-}C_6H_4\text{-}CH_2\text{-}C(CO\text{-}O\text{-}CH_2\text{-}CH_3)_2\text{-}NH\text{-}CO\text{-}CH_3 \xrightarrow{H^+} m\text{-}I\text{-}C_6H_4\text{-}CH_2\text{-}CH(COOH)\text{-}NH_2 \qquad (5.10)$$

Method

m-*Iodobenzyl bromide. m*-Iodotoluene (100 g) was placed in a 500 ml three-necked flask equipped with a mercury-sealed stirrer, a dropping funnel, and a reflux condenser to which a gas trap was connected. This arrangement was placed in an oil bath at 120 °C and illuminated with a 100 W tungsten lamp. The lamp, which was fitted with a reflector, was mounted approximately 20 cm from the flask. Bromine (27 ml) was added dropwise with constant agitation over 4 h. The reaction was stirred for an extra hour and left overnight at room temperature. Most of the Br_2 and HBr still present were then removed on the rotary evaporator. Petroleum ether, 40–60 °C (500 ml), was added to the liquid product and the vessel was cooled in a bath of isopropanol/solid CO_2. Solid material began to precipitate after the walls of the vessel were scratched with a glass rod. After 1 h, the solid was separated by decantation and suspended in 100 ml of fresh petroleum ether. The suspension was heated in a water bath at 50 °C until the solid dissolved. A small quantity of tar remained which stuck to the glass when the solution was transferred to a new vessel. The clear solution was cooled once again in the bath of isopropanol/solid CO_2 and solid collected as before. The cycle of solution at 50 °C followed by chilling was repeated two or three times more, using 100 ml of the petroleum ether each time. The solution, which had been black-violet in colour to begin with, became light red. Finally, the solid was washed with 100 ml of petroleum ether which had been pre-cooled in the bath. The yield of slightly pink solid (m.p. 43–46 °C) was 26.3 g.

DL-m-*Iodophenylalanine*. Sodium (1.01 g) was dissolved in 354 ml of absolutely dry ethanol in a 1 litre three-necked flask fitted with a mercury-sealed stirrer and a reflux condenser. Diethyl acetamidomalonate (9.61 g) was added and the mixture stirred for 1 h. *m*-Iodobenzyl bromide (13.15 g) was then added and the mixture refluxed for 5 h. About 95% of the ethanol was removed on the rotary evaporator; 100 ml of water were added to the concentrated mixture and extracted twice with 200 ml lots of peroxide-free diethyl ether, and twice with 100 ml lots. The combined ether layers were rotary evaporated to a brown oil. Hydrobromic acid (47%, w/v in water; 150 ml) was added to this oil and refluxed for 22 h. The acid was removed on the rotary evaporator. The dry solid was dissolved in 100 ml of water and filtered. The pH of the solution was taken to 6.5 with NH_3 solution (25%, w/v). A copious precipitate formed at once and, after 3 h at 4 °C, the solid was filtered off under suction. The solid was then suspended in 100 ml of 50% (w/v) ethanol. The apparent pH (glass electrode) was adjusted to 2.0 with 6 M HCl (the solid then dissolved) and then to 6.5 with NH_3 solution (25%, w/v). The mixture was left at 4 °C for 24 h and the solid filtered and recrystallized by the same procedure. The yield of the DL-amino acid (m.p. 213–217 °C, lit. 213.5–216.5 °C) was 5 g.

Notes

(1) The preparation of the iodobenzyl bromide is that of Weitzman and Patai (1946). Borrás (1972) found that, notwithstanding the report of Counsell *et al.* (1970), the method would not in his hands produce the analogous *o*-iodo product. Borrás obtained the *o*-iodo product by the method of Mabery and Robinson (1882) exactly as described by those authors except for the substitution of powdered ice from the refrigerator rather than the freshly fallen snow that they recommended. Borrás obtained 1.88 g of product from 26 g of *o*-iodotoluene, but found difficulties in scaling up.

The difficulties in making the *ortho* compound are presumably due to steric factors.

(2) The positioning of the lamp at 20 cm from the flask was found to be critical.

(3) It will be appreciated that the major difficulties in this Example are those of obtaining the halide with which to carry out the alkylation. The subsequent condensation is normally quite straightforward.

(4) The DL-mixture can be resolved according to Example (5.10) and converted to an —NH_2-protected active ester by the method of Example (5.5).

Example (5.9). The resolution of the DL-fluorophenylalanines

Source. Saunders (1974) based on Bennett and Niemann (1950), Waldschmidt-Lutz and Kühn (1951), and Milne and Most (1968).

Process involved. Reactions (5.2) and (5.3).

Method

The *o*-, *m*-, or *p*-fluorophenylalanine was given N^{α}-protection by the Boc-group by the method of Example (5.5). The product was dissolved (at a concentration of approximately 30 mg/ml) in 0.5 M sodium acetate, pH 6.0/0.025 M cysteine hydrochloride. An equivalent amount of phenylhydrazine (0.37 ml/g of Boc-amino acid) was stirred into the solution. Crude papain powder (type II, Sigma Ltd, approx 2 i.u./mg protein) was added (one-quarter of the weight of the Boc-amino acid) and the solution readjusted to pH 6 with dilute NaOH or acetic acid as necessary. The solution was incubated for 16 h at 37 °C in stoppered centrifuge tubes. The precipitate was collected by filtration and washed with water (crop 1). The materials were recrystallized from ethyl acetate/hexane and then twice from methanol/water. The yields (based on the content of the L-form in the original mixture), m.p., and $[\alpha]_D^{20}$ from a 1% (w/v) solution in methanol were: *o*-, 64%, 150–151 °C, −17.3°; *p*-, 52%, 133–134 °C, −17.1°. The crystals were needle shaped. The m.p. of the *meta* derivative was 141 °C.

The filtrate from crop 1 was readjusted to pH 6 by the addition of dilute acid and incubated for a further 24 h. A precipitate was formed and was collected and recrystallized as before. Yields (based on the quantity of the L-form in the original mixture), m.p., and $[\alpha]_D^{20}$ from a 1% (w/v) solution in methanol were: *o*-, 20%, 181–182 °C, −0.3°; *p*-, 30%, 172–173 °C, −1.0°. The crystals were plate shaped. The m.p. of the *meta* compound was about 162 °C. The filtrate from crop 2 was adjusted (without any precipitate being formed) to pH 5.0 and the incubation continued for 44 h. The precipitate was collected and recrystallized as before. The data for the products were: *o*-, 50%, 150–151 °C, +17.4°; *p*-, 30%, 130–131 °C, +18.1°. The crystals were needle shaped.

The phenylhydrazides were then dissolved or suspended in water (approximately 20 mg/ml) and an equal volume of 1 M $K_3[Fe(CN)_6]$ added. The mixture was refluxed for 2 h. The removal of the phenylhydrazide group was followed by the acidification of samples of the reaction mixture with solid citric acid, extraction of the amino-acid derivatives into ethyl acetate, drying, treating with anhydrous trifluoroacetic acid for a few minutes, drying again and electrophoresing on paper at pH 1.9. The free amino acids ran with a mobility of 0.64 with respect to that of serine while the phenylhydrazides ran with a mobility of 0.95. When the phenylhydrazide spot had disappeared (approximately 2 h), the neutral reaction solution was extracted with ethyl acetate. The aqueous layer was then adjusted to pH 3 with solid citric acid and again extracted with ethyl acetate. The ethyl acetate layer was dried over anhydrous Na_2SO_4 and then taken down to a solid by rotary evaporation. Yields of 80–90% were obtained, based on the quantity of phenylhydrazide that was taken to begin with.

Notes

(1) The racemic fluorophenylalanines are commercially available.

(2) The papain, Sigma type II, is the dried latex of the plant, freed from insoluble material but otherwise unfractionated. We have found it perfectly satisfactory and much cheaper per enzyme unit than the more purified grades of the enzyme. If it was not perfectly soluble to begin with, it had to be clarified by centrifugation.

(3) It appears that crop 1 is the L-form of the phenylhydrazide, crop 2 a mixture, and crop 3 the D-form. Care must be taken to control the process: the proper time for the collection of the L-form will presumably vary from one amino acid to another.

(4) The synthesis of the phenylhydrazide had to take place, particularly at the start, at a pH somewhat higher than the optimum for the enzyme (which is approximately 4.6) because of the insolubility of the Boc-amino acid at lower pH's. The slower rate of synthesis consequent on this fact may have contributed to the successful distinction between the D- and L-forms.

(5) The phenylhydrazides had $R_f = 0.68$ in the thin-layer system of Example (A.9), and stained orange with the ninhydrin reagent described in Example (A.4). Some of what was probably a salt of the Boc-amino acid was present in each of the three crops ($R_f = 0.34$) but was removed by the recrystallization.

(6) Cu^{2+} was found less effective than $[Fe(CN)_6]^{3+}$ for the oxidation of the phenylhydrazides.

(7) The electropherograms were stained with cadmium–ninhydrin (Example (A.4)). Samples that had not been treated with trifluoroacetic acid before running did not show any colour with 1% ninhydrin in acetone, and so it was concluded that the Boc- group survived the oxidation. (The cadmium–ninhydrin reagent could not be used for the latter test, as the acetic acid that it contains can partially remove the Boc- group and then give a positive test for free—NH_2.)

(8) All the test electrophoreses contained another spot, mobility 1.2, which was assumed to be the di-N^{α}-Boc-aminoacyl-phenylhydrazide, as reported by Milne *et al.* (1957). It was not susceptible to oxidation.

(9) The extraction of the oxidized product at neutral pH removed any remaining phenylhydrazide and the di-aminoacyl-phenylhydrazide.

(10) The method has the advantage that the N^{α}-protection, which is necessary to satisfy the specificity requirements of the enzyme, is one of use in peptide synthesis. Also, it is not removed from the L-form during the resolution. These two features do not apply to Example (5.10) below although Greenstein (1957) recommends that method in preference to the present one for aromatic amino acids.

(11) The method has the apparent advantage that the phenylhydrazide group, rather than being removed, could be used as a form of pre-activation for coupling (Milne and Carpenter, 1968). Saunders (1974) attempted to activate the phenylhydrazide to the phenyldiimide by oxidation by *N*-bromosuccinimide but found that the coupling efficiency was not good and that a number of unknown protein side-products were formed.

Example (5.10). The resolution of DL-[3-^{13}C]phenylalanine

Source. D. J. Saunders and R. E. Offord (unpublished work), based on the methods of Vine *et al.* (1973), Schnabel (1967), and Weygand *et al.* (1956).

Process involved. The formation of the *N*-trifluoroacetyl derivative of the racemic mixture, followed by the removal of the trifluoroacetyl group from the L-form by carboxypeptidase A.

Method

Formation of the trifluoroacetyl derivative. DL-[3-^{13}C]Phenylalanine (332 mg, 90.0 atom % ^{13}C) was dissolved in 2.4 ml trifluoroacetic acid and cooled to 12 °C. Trifluoroacetic anhydride (0.32 ml) was added and allowed to react for 3 h at 8 °C. The mixture was then dried *in vacuo*, dissolved in methanol, and dried down again. The solid was taken up in 10 ml ether and the insoluble material removed by centrifugation. The supernatant was dried and the solid residue recrystallized from benzene/petroleum ether, 60–80 °C. Yield 83 %, m.p. 119–123 °C (lit. 126–127 °C for the ^{12}C compound).

Stereoselective cleavage by carboxypeptidase. The trifluoroacetyl derivative was dissolved in 25 ml water and the pH was adjusted to 7 by the careful addition of 2 M LiOH. A suspension of carboxypeptidase A (1 mg, 50 i.u., in 50 μl water) was added and stirred for 4 h at 37 °C. The enzyme was then denatured by heating for 45 min at 70 °C and removed by filtration. The mixture was acidified by the addition of an excess of 12 M HCl and extracted four times with a total of 60 ml of chloroform. The chloroform layer was dried over anhydrous Na_2SO_4, filtered, and dried. The aqueous layer was dried *in vacuo* and the resulting solid was dissolved in 96 % ethanol. The amino acid was precipitated from the solution (which also contained LiCl) by the addition of ether. The yield of trifluoroacetyl material from the chloroform layer was 216 mg, which corresponds to almost exactly the maximum yield based on the amount of the D-amino acid in the original sample. The $[\alpha]_D^{20}$ of N^α-trifluoroacetyl-D-phenylalanine was $-17.0°$, conc. = 2% in ethanol; lit. $-17.2°$, conc. = 2% in ethanol (Fones and Lee, 1954).

The L-fraction was used without further purification for the preparation of the Boc- derivative. It was dissolved in 4 ml of dioxan (peroxide-free)/water, and t-butyloxycarbonyl azide was added (0.264 ml, roughly a 2.2-fold molar excess). The externally indicated pH (p. 217) was adjusted to 10.1 with a few drops of 4 M NaOH and the pH was controlled thereafter by the addition of 1 M NaOH from an autotitrator. After 7 h, water (approximately 10 ml) was added and the solution was extracted with two 15 ml lots of ether. The aqueous layer was acidified with approximately 2 ml of 5% (w/v) $KHSO_4$. The milky suspension was extracted with four more 15 ml lots of ether and then with two of ethyl acetate. The pooled organic layers were dried over anhydrous Na_2SO_4 and taken down to an oil *in vacuo*. The weight of the oil (200 mg) was 90% of the theoretical yield, based on the weight of the L-amino acid in the original sample.

Formation of the trichlorophenyl ester. The oil was dissolved in 6 ml of ethyl acetate. A sample (0.66 ml, approximately 83 μmol) was taken and the rest of the solution was dried down. To the sample solution was added 2,4,6-trichlorophenol (18 mg, 87 μmol) and dicyclohexylcarbodiimide (19.7 mg, 90 μmol). The solution was stirred for 16 h at 8 °C. Examination of a small sample by the thin-layer system of Example (A.9) showed that the majority of the material had the same R_f as an authentic sample of the ^{12}C form of the ester. After 5 h more at 20 °C, the precipitate was removed by centrifugation and washed with ethyl acetate. The combined supernatants were dried *in vacuo*.

Notes

(1) The purpose of adding and then removing the methanol is to help drive off the last traces of trifluoroacetic acid and its anhydride.

(2) The trifluoroacetyl material was only recrystallized once, and the free L-amino acid, its Boc- derivative and the active ester not at all, because of the desire to avoid losses on handling. (A trial recrystallization of the Boc- material from ether/petroleum ether in the cold was unsuccessful.)

The main impurity in the L-amino acid is a trace of LiCl. Lithium chloride does not interfere with subsequent operations and is removed during them. The general technique of the recrystallization of the trifluoroacetyl derivative was similar to that of Example (5.6).

(3) The carboxypeptidase A was of the diisopropyl fluorophosphate-treated type.

(4) The trifluoroacetyl material recovered from the chloroform layer is largely the D-form.

(5) The quantities taken for the $[\alpha]_D^{20}$ measurements could be estimated by the amino-acid analyser.

(6) We found the 2,4,6-trichlorophenyl ester of the Boc-L-[3-^{13}C]phenylalanine coupled quite satisfactorily in subsequent experiments.

(7) The trifluoroacetylation step is based on the method of Weygand *et al.* (1956). Greenstein (1957) recommends carboxypeptidase A for the resolution of the aromatic amino acids and the papain method (Example (5.9)) for most other amino acids (see Note 10 to Example (5.9)). The $-NH_2$ protection in Greenstein's method is by the chloroacetyl group. When working with the fluorophenylalanines, D. J. Saunders (unpublished work) found the trifluoroacetyl group was more convenient to remove than the chloroacetyl group.

(8) The dioxan was passed through an alumina column before use to remove peroxides.

(9) The 2.2-fold excess of the azide was originally used by mistake: a 1.1-fold excess was recommended by Schnabel.

(10) The trifluoroacetyl group was removed from the D-fraction by hydrolysis in a sealed tube with 12 M HCl (British Drug Houses Aristar grade) at 108 °C for 24 h. After the sample had been dried down, and several times redissolved in water and redried, the solid was given Boc- protection as above (but using a 1.1-fold molar excess). Yield 70%.

Example (5.11). A simple route to novel L-amino acids

Source. Moore (1978) and R. R. Moore and R. E. Offord (unpublished work).

Process involved. Reaction (5.5); Reaction (5.6). The process works equally well with asparagine or glutamine. The details below refer to glutamine.

Method

N^{α}-Boc-L-glutamine was purchased, or made by the method of one of the preceding Examples. It was dissolved in 3 M NaOH to a concentration of 2 mol/l. To 1 ml of this was added 2.5 ml of 0.8 M Br_2 in 3 M NaOH, both solutions having been cooled to 4 °C. The mixture was left for a few minutes until the yellow colour was discharged. The solution was heated at 75 °C for 30 min. The reaction mixture was then cooled and neutralized with 3 M HCl. It is essential to add the acid in very small increments, and to stir vigorously, in order to prevent the Boc- groups being removed by locally high concentrations of HCl. The neutral solution was diluted with water until approximately 10 mg/ml in amino acid. The product was desalted with AG11A8 resin (Bio-Rad Laboratories). A column 2.5 cm × 90 cm (450 g of resin) gave good resolution with a loading of up to 40 ml of such a solution. The eluant was water, and the flow rate 360 ± 120 ml/h. The amino acid was detected in the effluent by electrophoresis of small samples. The peak fractions were dried. If desired, the product can be recrystallized from methanol/acetone.

The new γ-NH_2 group was then substituted by a fluorescent group in the following manner. The material from the column was made up as a 2 mg/ml solution in water/ethanol (1 : 9, v/v) and mixed with an equal volume of 2 mg/ml solution of 1-nitro-4-chlorobenz-2,3-furazan in absolute ethanol. To 19 vol. of this solution was added 1 vol. of ethanol that had been saturated with sodium acetate trihydrate. The mixture was heated at 70 °C for 6 h and then rotary evaporated to dryness. The product was purified by chromatography on a column (5 cm × 75 cm) of polycaprolactam power in ethyl acetate/acetic acid/ methanol, 20 : 1 : 1 (by volume). The column was packed in 800 ml eluant and washed with a further 1 litre. The sample (200 mg) was applied in 10 ml of eluant. The column was eluted at about 400 ml/h. The major peak of absorbance at 450 nm eluted between 1.5 and 2.7 litres and proved to be the product. The pooled fractions were washed repeatedly with water until the aqueous washings were neutral. The ethyl acetate layer was dried with anhydrous $MgSO_4$ and rotary evaporated to an oil.

Notes

(1) The active ester of the product can be made by the methods of preceding Examples, using ethyl acetate as a solvent.

(2) It is surprising that the very basic conditions of the first part of the Method do not lead to racemization, but polarimetry shows that they do not.

(3) The condition of the Hofmann degradation are critical, and should not be altered.

(4) Direct sunlight should be avoided when working with the furazan and its derivatives.

CHAPTER 6

Coupling

We have now reached the position that we have striven so long to achieve. If we have done our work properly, we should now possess, in their proper state of protection, all the fragments that we need to make up our protein analogue. They will differ rather little from the protected fragments made by conventional synthesis except that they are, as we have seen, likely to be somewhat more polar. With this similarity to fragments synthesized by conventional means, it is not surprising to find that the preferred methods of coupling are those that we discussed when considering conventional methods in Chapter 1. The azide, carbodiimide, and active-ester methods have all been used in semisynthesis. The strategy that we now need to adopt—whether we are to use stepwise coupling or fragment condensation—will depend on decisions taken long before.

STEPWISE COUPLING

The majority of published semisyntheses (see Chapter 8) have used the stepwise strategy. Active esters of N^{α}-protected amino acids are reacted in turn with the α-NH_2 of the truncated, ε-NH_2-protected molecule. The N^{α}-protection is chosen to allow the new α-NH_2 group to be liberated after each coupling by a specific deprotection reaction, and the cycles of coupling and deprotection of the α-NH_2 are continued until we have the desired molecule. These operations recall the well-known, but rather less closely controlled process in which amino acids are coupled simultaneously to all the —NH_2 groups (both α- and ε-) of an unprotected, native protein (e.g. Levy and Carpenter, 1967).

It is usual to ensure the completion of the coupling by adding a large excess of the active ester. Furthermore, König and Geiger (1972) have shown that, in certain solvents, the presence of free 1-hydroxybenzotriazole enhances the rate at which some active ester couplings proceed. This molecule, which can itself form active esters (Table 1.4), is found to catalyse coupling to the α-NH_2 group of valine B2 of N^{α}-Gly^{A1},N^{ε}-Lys^{B29}-di-Boc-des-Phe^{B1}-insulin. Coupling at this site appears to suffer from steric hindrance and the considerable degree of acceleration given by the 1-hydroxybenzotriazole is of real value. Example (6.1) gives the standard experimental conditions used in this Laboratory for the addition of a single residue to a truncated chain. The conditions do not differ greatly from those used by others. Example (6.2) describes the stepwise addition of more than one residue—the process described is the first part of the semisynthesis of the modified fragment of cytochrome *c* discussed on p. 184. Example (6.3) indicates the additional precautions necessary for a stepwise semisynthesis at the radiochemical level.

FRAGMENT CONDENSATION

Fragment condensation has so far been limited to the coupling of tryptic and CNBr fragments. Both types of cleavage have, as we have seen, special advantages for the subsequent discrimination between α- and side-chain —COOH groups.

The use of tryptic peptides

Offord (1969a) proposed the scheme for fragment-condensation semisynthesis with tryptic peptides that is set out in the preceding Chapters, and reported the coupling, in poor yield, of two tetrapeptides. Izumiya *et al.* (1971) soon showed, using synthetic peptides of appropriate sequence, that there seemed to be no bar to the semisynthetic coupling of tryptic peptides in high yield. Yet Smyth (1975), in a review of polypeptide semisynthesis, rightly remarked that 'while these reactions [the scheme proposed in 1969] seem straightforward, it is true to say that few reports have appeared in the literature on the preparation of new polypeptides by the combination of natural fragments'. In fact between 1969 and 1975 only three fragment-condensation semisyntheses were reported. Now that such preparations are being carried out in greater numbers (see Chapter 8) is it possible to say what was the cause of the delay? Each of those involved must supply their own answer. Mine is, in part, that the conversion of the other operations of conventional synthesis, and those of sequence determination, into the forms now proposed in Chapters 2–5 was not a trivial exercise. It was only natural that most early semisyntheses should be of the simpler, stepwise type. Then, when methods of protection and cleavage had been developed to the extent that serious attention could be given to the coupling problem, it became apparent that the efficiency of coupling of two natural fragments tended to be markedly lower than an otherwise closely similar coupling between synthetic fragments. This difficulty was almost certainly due to contamination of the peptides by materials picked up during their preparation and isolation, and must now be considered.

The problem of contamination

Such traces of carbohydrate that might result from the preparation of the fragments on paper or on columns of cellulose or dextran do not seem to interfere with the coupling. On the other hand, traces of formic, acetic, and other carboxylic acids from the buffers commonly used in biochemical separation methods almost certainly do interfere and must be eliminated. It also seems prudent to guard against there being any traces of extraneous amino compounds

—such compounds are commonly present in the air and water of a biochemical laboratory. Some of the organic solvents that are commonly used give traces of amino and carboxyl compounds on decomposition, for example dimethylformamide can give rise to formic acid. Atherton *et al.* (1975) have shown that traces of formic acid can seriously interfere with couplings, and so it is wise only to use such a solvent immediately after purification. (The option also exists of replacing dimethylformamide by dimethylacetamide, on the grounds that the acetic acid arising from the latter is likely to be less objectionable, but we have not so far found this to be necessary in this Laboratory.)

The rationale behind Examples (6.4) and (6.5), which describe the final preparation of the protected amino and carboxyl components for coupling, is to assume that extraneous carboxyl and amino components are present and to seek to remove them either by rendering them volatile or by some technique such as gel filtration. Whether or not these contaminants are always present in harmful amounts, couplings between materials treated as these Examples describe are often much more successful than those in which even a minor detail is omitted.

Coupling

Couplings between tryptic peptides in which one is small and the other large have become fairly easy, particularly when material is spared by performing the preliminary trials on a scale that is small by biochemical, let alone organic-chemical, standards (Example (6.6)). The Example goes on to illustrate the preparation of a number of polypeptides from tryptic fragments, the longest product being of 51 residues. 'Easy' does not mean 'satisfactory in every detail'—for instance, it would be better to be able to use smaller excesses of carboxyl component. The carboxyl component has to be pre-activated, and, particularly when it is to be in excess, it is essential to ensure that activation has gone to completion before mixing with the amino component. If the period allowed for activation is too short, some carbodiimide will remain when the amino and carboxyl components are mixed. A quantity of unreacted carbodiimide that is small in relation to the carboxyl component could be, if the carboxyl component were in large excess, a large quantity in relation to the amino component. Side-reactions might then follow, particularly if the —COOH groups of the amino component were free. The lack of —COOH protection in couplings like those in Example (6.6), while permissible in principle, therefore carries some risk.

A peptide active ester remains liable to racemization from the time that it is formed until it participates in a coupling reaction—the risk is much greater than for a urethan-protected amino acid, cf. Reaction (1.9b). However, the low pH of the reaction medium for the activation suppresses this tendency to a great extent. The main risk arises at the coupling stage, which takes place under more basic conditions and therefore presents a more favourable opportunity for racemization to occur. In spite of this possibility (which is an equally serious risk in couplings between synthetic fragments), Rees and Offord (1976b) found no measurable racemization under the conditions of Example (6.6). Other conditions that they tested gave quite high degrees of racemization.

Couplings when both fragments are large tend to be difficult. The problem (but not the compensating possibility of its relief by steric assistance (p. 29)) is seen in conventional synthesis also, as we remarked in an earlier Chapter. I believe that more attention should be given to coupling large components in aqueous solution where there is the best opportunity to obtain steric assistance. It would be possible to use a water-soluble carbodiimide *in situ*, but then the —COOH groups of the amino component would require protection. In order not to depress solubility and to frustrate as little as possible the tendency to steric assistance, the —COOH protection should preferably be of the sort that leaves the ester with a negative charge (Example (4.6)). If we have to use a neutral ester, it should be the methyl, since this has least effect on solubility. The acetimidyl group is, for the same reasons, excellent for —NH_2 protection (Wallace and Offord, 1979).

There is no reason why natural fragments should not be coupled sequentially on a solid support, and Example (6.7) describes how one might set about doing so.

Cyanogen bromide fragments

We have seen in Examples (5.3), (5.5), and (6.2) a detailed description of how we can remove the homoserine residue from the carboxyl terminus of a CNBr fragment and prepare for its replacement by methionine in the final, resynthesized product. Example (6.8) completes the process. Couplings of this sort are no different in principle from those between tryptic fragments, except that when coupling tryptic peptides we are used to having to activate carboxyl-terminal arginine, but here we are just as likely to meet any of the other protein amino acids in the carboxyl-terminal position. We can even activate carboxyl-terminal homoserine if its presence in the final product is not objectionable (Wallace, 1976).

If the replacement of methionine by homoserine is tolerable, we have two other possibilities. Firstly, we can use the lactone form in which the homoserine is liberated by the CNBr cleavage (Reaction (3.5)) as an active ester, and allow coupling to occur directly on aminolysis by the amino component (Reaction (6.1)).

The lactone is not a very highly activated form of the —COOH group and its aminolysis tends to be slow. Nonetheless it has been used successfully several times, in particular on fragments of cytochrome *c* and of pancreatic trypsin inhibitor, using both modified and unmodified fragments (see Chapter 8). Some of these couplings may be instances of the phenomenon of steric assistance which we have considered before. Example (6.9) describes a condensation in which one of the fragments was modified and both were partially protected.

Offord (1972) proposed the use of the lactone as a temporary selective protection before carbodiimide-mediated couplings and, as an alternative, the use of the lactone as a specific activation for an azide coupling (Reactions (6.2) and

$$R_1\text{—NH—[COOH, NH—R}_2\text{, COOH]—NH—CH—C(=O)—O (lactone ring: } H_2C\text{—}C H_2\text{—O)} + H_2N\text{—[NH—R}_2\text{, COOH]—COOH}$$

$$\downarrow$$

$$R_1\text{—NH—[COOH, NH—R}_2\text{, COOH]—NH—CH(CH}_2\text{—CH}_2\text{OH)—CO—NH—[NH—R}_2\text{, COOH]—COOH} \quad (6.1)$$

(6.2a). Like Reaction (6.1), this method has the attraction that protection of the side-chain —COOH group is unnecessary.

Example (6.10) gives conditions, but there seems in practice to be little advantage over Reaction (6.1) or even the carbodiimide couplings. Wallace

$$R_1\text{—NH—[COOH, NH—R}_2\text{, COOH]—NH—CH—C(=O)—O (lactone ring: } H_2C\text{—}C H_2\text{—O)} + H_2N\text{—}NH_2$$

$$\downarrow$$

$$R_1\text{—NH—[COOH, NH—R}_2\text{, COOH]—NH—CH(CH}_2\text{—CH}_2\text{OH)—CO—NH—NH}_2$$

$$\downarrow HNO_2$$

$$R_1\text{—NH—[COOH, NH—R}_2\text{, COOH]—NH—CH(CH}_2\text{—CH}_2\text{OH)—CO—N}_3 \quad (6.2)$$

$$R_1\text{—NH—[COOH, NH—R}_2\text{, COOH]—NH—CH(CH}_2\text{—CH}_2\text{OH)—CO—N}_3 + H_2N\text{—[NH—R}_2\text{, COOH]—COOH}$$

$$\downarrow$$

$$R_1\text{—NH—[COOH, NH—R}_2\text{, COOH]—NH—CH(CH}_2\text{—CH}_2\text{OH)—CO—NH—[NH—R}_2\text{, COOH]—COOH} \quad (6.2a)$$

(1976) reports that the reversion of the peptide azide to the lactone is a major side-reaction—Reaction (1.5), the expected side-reaction, was not detected.

ENZYMES AS COUPLING AGENTS

Since many of the fragments that we use are produced by proteolytic enzymes, can the enzymes be used to promote the back reaction? Under normal experimental conditions the formation of a peptide bond is a rather unfavourable process from the point of view of thermodynamics, and it is necessary to use some special manœuvre to ensure a reasonable yield of product. Laskowski (1978) has listed the possible approaches, most of which involve shifting the position of equilibrium. First one can simply use mass action and drive the reaction in the direction of coupling by having a high concentration of one of the reactants. This approach, which has also been recently discussed with reference to conventional synthesis (Isowa *et al.*, 1977; Morihara and Oka, 1977), includes in addition those preparations in which the reaction goes in the direction of synthesis by virtue of the insolubility of the products, as in Example (5.9). Second, the equilibrium constant itself can sometimes be unusually favourable toward coupling as a result of the conformation of the product. This is the basis of the many semisyntheses of analogues of protease inhibitors, reviewed in Chapter 8. The third approach is to couple the equilibrium to a thermodynamically more favourable reaction: this too has so far found its application in the semisynthesis of analogues of protease inhibitors, and is also discussed in Chapter 8, together with a method that employs a *kinetic* rather than thermodynamic means to influence the direction of reaction. The last method of altering the position of equilibrium involves the use of organic solvents in the reaction mixture. Solvents that are neutral, non-ampholitic, and hydrophilic—such as glycerol and 1,4-butanediol—can, when present in high concentrations, alter the equilibrium constant very significantly in favour of synthesis. In 1,4-butanediol : water (4 : 1, v/v) the equilibrium constant is 100 times more favourable than in water (Homandberg *et al.*, 1978). Although these solvents are chosen for their limited ability to denature proteins, at these high concentrations they nonetheless reduce the catalytic efficiency of proteases. It may take a very long time to reach equilibrium which, even when it is attained, is unlikely to lie so very close to 100% synthesis. However, these potential drawbacks are minor in comparison to the attractiveness of a route that virtually excludes the need for side-chain protection. Chemical couplings are often slow, and far from complete, and it is to be hoped that it will prove possible to exploit this promising suggestion in practice. An important first step has recently been taken: Homandberg and Laskowski (1978, 1979) report that this approach had made it possible to restore the peptide link between the amino terminus of pancreatic ribonuclease S-protein and the carboxyl terminus of the S-peptide. In this instance the use of glycerol rather than butanediol led to success.

EXAMPLES

Example (6.1). The stepwise semisynthesis of B1 analogues of insulin

Source. Method A: Saunders (1974). Method B: Saunders and Offord (1977a). See Chapter 8 for a review of the many other similar syntheses.

Process involved. Reaction (1.10). The active ester in Method A is formed by *N*-hydroxysuccinimide. The active ester in Method B is formed by 2,4,6-trichlorophenol. The amino component is N^{α}-Gly^{A1}-,N^{ϵ}-Lys^{B29}-diacyl-des-Phe-insulin.

Method A

The diacyl-des-Phe-insulin (prepared by one cycle of the operation described in Note 3 to Example (3.1), or by one cycle of the method of Example (3.2)) was dissolved (2 mmol/l) in redist. dimethylformamide (p. 218). Solid Boc-amino acid active ester (20–40 μmol/ml of protein solution, prepared by the Methods given in Chapter 5) together with an equivalent amount of 1-hydroxybenzotriazole, was added and stirred at 20 °C. The externally indicated pH (p. 217) was adjusted to 8.5–9 with *N*-ethylmorpholine. When a check by the method of Examples (A.7), (A.8), (A.9), or (A.10) indicated that the insulin was behaving substantially as a triacyl derivative (usually after 16–24 h), the protein was precipitated by the addition of about 7 vol. of peroxide-free ether. The precipitate was collected by centrifugation and carefully washed with more ether (see Note 2 to Example (3.1)) and dried *in vacuo*. The product was deprotected by Method A of Example (7.1) or the Method of Example (7.7) as appropriate. The product was checked by the usual analytical tests, by crystallization (Example (7.10)) and by biological and immunological assay.

Method B

N^{α}-Gly^{A1},N^{ϵ}-Lys^{B29}-di-Boc-des-Phe^{B1}-insulin (50.6 mg, prepared as in Note 3, Example (3.1)), dissolved in 5 ml of freshly distilled dimethylformamide (p. 000), was placed in a stoppered test tube with Boc-L-[3-^{13}C]phenylalanine trichlorophenyl ester (an approximately 10-fold molar excess over free protein —NH_2 groups, prepared as described in Example (5.10)). The externally indicated pH (p. 217) was adjusted to about 8.5 with *N*-ethylmorpholine and the mixture was stirred for 20 h at 20 °C. The protein was precipitated after a further 6 h with approximately 7 vol. ether. The precipitate was collected by centrifugation and washed carefully with more ether. The combined supernatants were dried for recovery of ^{13}C material and the protein pellet dried *in vacuo*. The protein (approximately 45 mg) was deprotected by Method A of Example (7.1) and was shown by the usual analytical tests and biological activity to be indistinguishable from native [^{12}C]insulin (D. J. Saunders and R. E. Offord, unpublished work). The test described in Example (A.10) indicated that coupling had gone substantially to completion.

Notes

(1) The addition of 1-hydroxybenzotriazole was recommended by König and Geiger (1972) as a means of accelerating the coupling at sites such as Val^{B2}, which is sterically hindered. Saunders (1974) found that, in agreement with the data given by König and Geiger, it reduced the time of a coupling of a very similar active ester to that above (of *o*-fluorophenylalanine) from 18 h to less than 1 h.

(2) See Note 4 to Example (6.6) for the appropriate molecular weight of the 1-hydroxybenzotriazole.

(3) Saunders (1974) and Saunders and Offord (1977a) report the semisynthesis of a number of B1 analogues of insulin, using active esters in molar excess of between 10- and 20-fold for times between 16 and 22 h. Borràs and Offord (1970a, b), Borrás (1972), and Saunders (1974) used trifluoroacetyl protection of the protein as well.

Example (6.2). The stepwise semisynthesis of an analogue of the CNBr fragment of cytochrome *c* corresponding to residues 81–104

Source. Wallace (1976) and Wallace and Offord (1979).

Process involved. Coupling, first of N^{α}-Boc-*o*-fluorophenylalanine, then after N^{α}-deprotection, of N^{α}-Boc-isoleucine, then, again after N^{α}-deprotection, of N^{α}-Boc-methionine to Ala-Gly-Ile-Lys(Acim)-Lys(Acim)-Lys(Acim)-Thr-Glu-Arg-Glu-Asp-Leu-Ile-Ala-Tyr-Leu-Lys(Acim)-Lys(Acim)-Ala-Thr-Asn-Glu. This peptide is the des-Ile-Phe derivative of the CNBr fragment of cytochrome *c* corresponding to residues 81–104. The couplings were effected via the *N*-hydroxy-succinimido esters.

Method

The peptide (10 μmol, prepared as in Example (5.2)) was dissolved in 10 ml of anhydrous dimethylformamide (p. 218). To the solution were added 10-fold molar excesses of the protected *o*-fluorophenylalanine active ester and of 1-hydroxybenzotriazole, both in the solid form. The externally indicated pH (p. 217) was adjusted to pH 8.5 with *N*-ethylmorpholine. The solution was allowed to stand for 24 h at 20 °C. Water (2 ml) was added to the reaction mixture (with cooling). The mixture was then extracted twice with ethyl acetate to remove unwanted materials and the more aqueous phase applied directly to a column of Sephadex LH-20 (2.5 cm × 100 cm) run in dimethylformamide/water (17:3, v/v). The peak eluting at the known position of the parent peptide was pooled and freeze dried. Recoveries of material were nearly 100 %. The efficiency of incorporation was checked by amino-acid analysis, end-group determination (Examples (A.1) and (A.2)) and (for the fluorophenylalanine step) nuclear magnetic resonance (Wallace, 1976). The α-NH_2 group was deprotected by dissolving in anhydrous trifluoroacetic acid (2 ml) for 1 h at 0 °C. The trifluoroacetic acid solution was then diluted fivefold with ice-cold ether and the precipitate spun down. The precipitate was washed in ether, with the precautions specified in Note 2 to Example (3.1), and dried under a gentle stream of N_2. It was then redissolved in 7 % formic acid and subjected to gel filtration on a column of Sephadex G-50 (135 cm × 2 cm). The peptide peak was collected and freeze dried. Isoleucine was coupled on by the same method except that trial experiments showed a 40–50-fold excess of active ester and 1-hydroxybenzotriazole to be necessary. Methionine was coupled on using a 30-fold excess of these two reagents. The final yield of product, checked as in Example (A.1) was 4 μmol.

Notes

(1) Other peptides may, in the separation immediately after the coupling, call for one of the alternative gel-filtration systems mentioned in Note 3 to Example (3.8). If solubility permits it is even possible to use 7 % formic acid on G-50 at this stage. Although this solvent will partly remove the Boc- group, this is not objectionable, providing the coupling has gone to completion.

(2) The overall yield is not corrected for losses during trial runs.

(3) The peptide is now ready to be coupled to the des-homoserine derivative of the CNBr peptide corresponding to residues 66–80 of the cytochrome (Example (5.3)).

Example (6.3). The semisynthesis of [Phe^{B1}-^{3}H]insulin at high specific activity

Source. Halban and Offord (1975).

Process involved. Reaction (1.10a), in which R = (succinimido ring: O, N—, O),

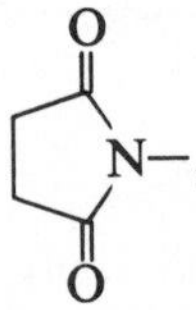

followed by Reaction (1.13) in which R = $(CH_3)_3C$—. Both processes take place at the μg level.

Method

The 1,2-dimethoxyethane solution of active ester from Example (5.7) was added to the N^{α}-Gly^{A1},N^{ϵ}-Lys^{B29}-di-Boc-des-Phe^{B1}-insulin from Example (3.1) in a 10-fold molar excess. The protein was dissolved at a concentration of 6 μg/μl in redist. dimethylformamide (p. 218). The reaction vessel was a Durham tube (6 mm internal diameter) drawn to a point. As soon as the solutions were mixed the 1,2-dimethoxyethane was removed by gentle blowing with N_2. If need be, the externally indicated pH (p. 217) was adjusted with *N*-ethylmorpholine so as to be above 8.5.

After 24 h at 20 °C the coupled product was precipitated by the addition of 5 vol. of peroxide-free ether, followed by centrifugation of the stoppered vessel in a bench centrifuge. The precipitate was washed with ether (employing the same precautions as those in Note 2 to Example (3.1)) and dried. The product was checked by isoelectric focussing (Example (A.8)) and deprotected by the method of Example (7.2).

Notes

(1) The dimethoxyethane was removed because of its unknown effect on the solubility and integrity of the insulin derivative; it could not be removed before mixing with the much less volatile dimethylformamide because of the decomposition that ensues if the ^{3}H-labelled active ester is dried down (see Note 5 to Example (5.7)).

(2) In spite of the very small quantities involved, the ether precipitation is very efficient. There is no measurable loss of protein and the small quantity of active ester that remains is readily removed at a later stage of the preparation (Example (7.2)).

(3) The product can be located after isoelectrophoresis by the use of a scanning detector equipped for windowless, gas flow counting. Because of shielding in the gel 50–100 μCi of product are required. The product migrated as a single, symmetrical peak in the position of a triacyl insulin.

Example (6.4). Fragment condensation: the pre-treatment of the amino component

Source. Rees and Offord (1976b).

Process involved. The conversion of $—NH_3^+$ to $—NH_2$ with simultaneous removal of volatile amino compounds.

Method A

The amino component was dissolved in dimethylformamide or, if it was insoluble in that solvent, dimethylsulphoxide. It was taken to pH 9–10 (externally indicated, p. 217) with triethylamine. If extensive contamination with volatile amino compounds was suspected the pH was taken close to 10, the solution dried, and the material redissolved in the original solvent.

Method B

If the peptide was soluble in water, it was dissolved (in 1 ml) and carefully adjusted to pH 10 with 1% (v/v) aqueous triethylamine. It was then passed down a column of Sephadex G-25 (1 cm × 30 cm) in 0.1 M aqueous pyridine and the peptide or protein peak freeze dried.

Notes

(1) Volatile —NH_2-containing materials derive from the buffers and solvents used in earlier stages, and possibly from the atmosphere of the laboratory as well.

(2) If Method B is used, steps must be taken (e.g. by a trial on a small quantity) to be sure that the material does not precipitate in the column. Also, since the 0.1 M pyridine will not overcome ionic interactions to any great extent, the possibility must be guarded against that a very basic peptide might be retarded by ion exchange with the negative charges that exist even on the nominally neutral Sephadexes.

(3) It is essential to avoid contamination via the vapour phase from other, —NH_2-containing samples on the freeze drier.

Example (6.5). Fragment condensation: the pre-treatment of the carboxyl component

Source. Rees and Offord (1976b).

Process involved. Conversion of —COO^- to —COOH with simultaneous removal of volatile carboxyl compounds.

Method

The carboxyl component was dissolved or triturated in 2–3 ml water cooled to 0 °C. Trituration of very intractable samples was assisted by dissolving them in the minimum volume of dimethylformamide or dimethylsulphoxide before adding the water. Sufficient 0.01 M HCl was then added to bring the pH (externally indicated on pH paper, using a 0.1 μl drop) to about 2.5. The sample was immediately freeze dried.

Notes

(1) As in Example (6.4), it is essential to avoid vapour-phase contamination on the freeze drier.

(2) Traces of formic acid are particularly objectionable and traces of acetic acid only a little less so. If present during the activation step (Example (6.6)), these two acids appear to be particularly prone to acylate the amino component (cf. p. 120). Since the usual preliminary tests for successful coupling involve the detection of any remaining free —NH_2 groups, the acylation contributes a spurious impression of success, which may not be dispelled until much later.

Example (6.6). The fragment condensation of protected tryptic peptides

Source. Offord (1973) and Rees and Offord (1976b).

Process involved. Coupling by the pre-activation of the carboxyl component to the 1-hydroxybenzotriazole ester.

$$\mathrm{R_1{-}NH{-}[COO{-}R_2,\ NH{-}R_3,\ CH_2CH_2CH_2NH{-}C(NH_2)(=NH_2^+)]{-}COOH} + \text{1-hydroxybenzotriazole} \xrightarrow{\text{carbodiimide}} \mathrm{R_1{-}NH{-}[COO{-}R_2,\ NH{-}R_3,\ CH_2CH_2CH_2NH{-}C(NH_2)(=NH_2^+)]{-}CO{-}O{-}N}(\text{benzotriazole}) \tag{6.3}$$

$$\mathrm{R_1{-}NH{-}[CO{-}O{-}R_2,\ NH{-}R_3,\ CH_2CH_2CH_2NH{-}C(NH_2)(=NH_2^+)]{-}CO{-}O{-}N}(\text{benzotriazole}) + \mathrm{H_2N{-}[NH{-}R_3,\ COOH]{-}COOH} \longrightarrow \mathrm{R_1{-}NH{-}[CO{-}O{-}R_2,\ NH{-}R_3,\ CH_2CH_2CH_2NH{-}C(NH_2)(=NH_2^+),\ NH{-}R_3,\ COOH]{-}COOH} + \text{1-hydroxybenzotriazole} \tag{6.3a}$$

It is essential to find the optimum lengths of time and temperatures for the activations (Scheme (6.3a)) by preliminary experiments on a small scale. The trial activation mixtures can be set to couple with practically any small, readily available peptide (in the present Example, the synthetic peptide Ser-Pro-Phe-Arg(NO_2)-benzyl ester) and then, as a check, with the amino component to be used in the bulk coupling.

Method

Stage 1. Determination of the correct conditions of activation of the carboxyl component. 1 μmol of carboxyl component (pre-treated as in Example (6.5)) was dispensed into a small Durham tube by dissolving the stock sample in methanol (usually 40 μmol/ml) and taking a measured sample. Both the stock and the sample were dried down with a gentle stream of N_2. 7 μl of dimethylformamide were added to the tube and the solution cooled to 0 °C. 2.5 μl of each of 0.44 M solutions (freshly made up in redist. dimethylformamide, p. 218) of 1-hydroxybenzotriazole and dicyclohexylcarbodiimide were then added. 3 μl samples were removed from the mixture at measured intervals, first at 0 °C and then (using the mixture that remains from 0 °C time course) at 20 °C. Each was immediately expressed from the microcapillary onto a Teflon sheet, mixed with 3 μl of a 16 mM solution of the synthetic peptide in dimethylformamide (previously adjusted to an externally indicated pH of 8.5 (p. 217) with *N*-ethylmorpholine) and then drawn up into the capillary again. The last 3 μl sample of the activated carboxyl component was mixed with the amino component in the Durham tube rather than being withdrawn. 0.1 μl of each mixture was taken to check that the externally indicated pH remained above 8. The capillary was then sealed at both ends with Parafilm and left for 24 h at 20 °C. The contents of each capillary were then expressed in two separate lots onto the origin line of a paper chromatogram. Whatman No. 4 paper was used because of its rapid running properties. 3 μl of the solution of the synthetic peptide were used as a marker. The chromatogram was developed for a few minutes (distance moved by front = 8 cm) with butanol/acetic acid/water/pyridine (15 : 3 : 12 : 10 by vol.; Waley and Watson, 1953) and stained with the Sakaguchi reagent for arginine and with Cd–ninhydrin for free amino component (Example (A.4)). There was usually a middle range of activation conditions in which the characteristic yellow colour given by the amino component had completely disappeared. Longer and shorter times led to there being some staining material left. The loss of colour was assumed to indicate complete coupling and the poorer results with the longer activation conditions were taken to be due to a rearrangement of the active ester to an inactive form (see Note 5). The longest of the satisfactory activation conditions were adopted.

The activation conditions were then checked by similar small-scale experiments using a solution of the appropriate natural amino component. The concentration and pre-treatment of the amino component were the same as those noted in Table (6.1) below. Dimethylsulphoxide was in general a more satisfactory solvent than dimethylformamide. As above, the ratio between carboxyl and amino component was 7 : 1. The optimum times and temperatures of activation were invariably the same for both the synthetic and the natural amino component.

Finally, a set of trial couplings was carried out with the natural amino component to establish the minimum ratio of the carboxyl component to amino component required to drive the coupling to completion in 24 h at 20 °C. Some couplings were found to require smaller ratios than 7 : 1 (see Table 6.1).

Table 6.1 Individual variations in the bulk couplings between tryptic fragments of lysozyme. (This Table is reproduced from Rees & Offord (1976b) by permission of the Biochemical Society.)

Coupling	Activation of carboxyl components	Quantity of carboxyl component used (μmol)	Quantity of amino component used (μmol) (volume in brackets, ml)	Solvent used for amino component	Pre-treatment of amino component (Example (6.4))
M7 to M8	60 min at 0 °C followed by 60 min at 20 °C	110	15.7 (1.22)	Dimethylformamide	Method B
M6 to M78	60 min at 0 °C	30	6 (0.2)	Dimethylsulphoxide	Method B
M3 to M4	60 min at 0 °C followed by 12 min at 20 °C	210	50 (2.4)	Dimethylsulphoxide	Method A
M2 to M34	35 min at 0 °C	110	20 (1.3)	Dimethylsulphoxide	Method A
M1 to M234	60 min at 0 °C	60	12 (0.5)	Dimethylsulphoxide	Method A

Notes

(1) The sequences of the peptides, M1, M2, etc. are given in Fig. 8.2.

(2) The activation mixtures of the carboxyl components were 83.3 mM in peptide except for M7 which was 0.1 M. The solvent was always dimethylformamide.

(3) Where gel filtration was used to remove carboxyl component from the coupling mixture before deprotection, the solvent was that used for the amino component prior to coupling.

Stage 2. Large-scale couplings. Large-scale couplings were carried out using the optimum conditions indicated by the preliminary experiments (details are given in Table 6.1). The mixtures were stirred magnetically. As a safeguard against accident, the couplings were carried out as two separate operations, each on half the indicated quantities. The progress of the reaction was followed by paper chromatography of small samples of the coupling mixture as before. As judged by the intensity of the ninhydrin stain, coupling was usually half complete after about 2–3 h and complete after 24 h. The material was then taken to dryness on a freeze drier and much of the excess carboxyl component and reagents extracted with three applications of 1 ml of methanol. The material that remained was then purified by gel filtration on a column. Peptides M78, M678, and M34 (the peptide names are those in Scheme (6.3)) were separated in the protected state on Sephadex LH-20 (105 cm × 1.4 cm). M78 and M678 were run in dimethylformamide, while M34 was run in dimethylsulphoxide. It was found desirable when using dimethylsulphoxide to pre-equilibrate the column with freshly distilled eluant. Peptide M234 did not separate from residual M2 on LH-20. The peptide was therefore deprotected with trifluoroacetic acid (Example (7.1), Method A) and run, 10 μmol at a time, on a column (105 cm × 1.4 cm) of G-25 in 0.025 M NH_4OH, brought to pH 10 with solid NH_4HCO_3. The column had been pre-treated to prevent adsorption by passing down 1 ml of a 1% solution (w/v) of bovine serum albumin. A trace of M2 still remained in the M234 and was removed as peptide M12 after the next coupling. Peptide M1234 would not separate from M1 and ultraviolet-absorbing by-products when run on the protected state on LH-20. Rees and Offord (1976b) report the special measures necessary for its purification, which involve a G-25 column run at a slightly elevated temperature.

Notes

(1) Although moderate quantities of natural fragments are on the whole easier to obtain than synthetic ones, and the effort does not increase very markedly with length, I find the use of the very small quantities in the trial couplings to be essential. The scale and the rapid means of assessment of efficiency provided by the chromatographic system make possible the exploration of a wide range of several variables without compromising the ability to carry out a bulk coupling afterward. With care the small-scale results are reproducible and can be applied with confidence to larger-scale work.

(2) The dispensing of the carboxyl component with methanol may have the incidental benefit of removing traces of moisture from the stock.

(3) *N*-Ethylmorpholine is the preferred base, for the reasons given in Note 1 to Example (2.1). *N*-Methylmorpholine is a slightly stronger base and can be used if, for any reason, the externally indicated pH has to be held between about 9.5 and 10. Pyridine acts best between 6.5 and 7 and collidine between 7 and 7.5.

(4) The correct molecular weight for the computation of the strengths of 1-hydroxybenzotriazole solutions is 153. This value takes into account one tightly bound molecule of water of crystallization. The water does not appear to interfere with the coupling.

(5) Electrophoreses of over-activated material indicate that the —COOH group is still substituted.

(6) Rees and Offord (1976b) could find no tendency to racemization when using those methods.

Example (6.7). The attachment of a tryptic peptide to a solid support

Source. W. H. Johnson (unpublished) adapted from the methods reported by Stewart and Young (1969). See the Supplement to Rees and Offord (1976a).

Process involved. The coupling, first of leucine and then the peptide Thr-Pro-Gly-Ser-Arg to a chloromethyl-polystyrene resin. The peptide is a tryptic fragment of hen-egg lysozyme.

Method

Boc-leucine (0.42 g) and triethylamine (0.18 g) were dissolved in 5 ml of absolute alcohol and added to 1 g chloromethyl-polystyrene resin (the resin contained approximately 2 mequiv. of Cl per g). The mixture was refluxed for 2 days. The resin was filtered off and washed with ethanol, water, acetic acid, methanol, dimethylformamide, with methanol once more, and dried *in vacuo*. The extent of incorporation of leucine was assessed by amino-acid analysis (Example (A.1)) of a 1 mg sample of resin. Approximately 400 μmol of leucine were incorporated per 1 g of resin.

The Boc- group was removed by transferring the resin to a stoppered tap funnel fitted with a sinter and shaking with the following reagents: dioxan (3 times); 4 M HCl (dry) in dioxan (once, for 30 min); dioxan (3 times); chloroform (3 times); triethylamine/chloroform (1 : 9, v/v) (once, for 10 min); chloroform (3 times); and dimethylformamide (3 times).

The leucyl-resin (1 mg) was treated with 7 μmol of the 1-hydroxybenzotriazole ester of the N^{α}-Boc-peptide, prepared as described in Example (6.6). Dimethylformamide was added until the total volume was 0.4 ml. After 2 days at 20 °C, the resin was washed 10 times with dimethylformamide. The coupled materials were removed from the resin by HBr/trifluoroacetic acid as described by Stewart and Young. Amino-acid analysis showed a 37% yield of hexapeptide, based on the quantity of leucine coupled to the resin.

Notes

(1) The method as described was a pilot study for the semisynthesis of a portion of hen-egg lysozyme on a resin support. The carboxyl-terminal residue of the protein is leucine, and it is released as the free amino acid by trypsin. For work on this protein, therefore, it is permissible to begin by coupling a free amino acid to the resin. A peptide might not have withstood the refluxing involved, and other, gentler methods would have to be found (probably involving the improved resins now being developed) to be able to begin the resin-bound semisynthesis of most proteins.

(2) 400 μmol of amino acid per 1 g of resin may be rather too much. The amount that does not react in the first coupling could be irreversibly protected with, say, acetic anhydride, or a lesser quantity of leucine coupled to the resin.

(3) The dilution of the coupling mixture may have slowed the rate of coupling somewhat.

(4) See also Fontana *et al.* (1971).

Example (6.8). The semisynthesis of an analogue of a portion of cytochrome *c* by fragment condensation

Source. Wallace (1976, 1978) and Wallace and Offord (1979).

Process involved. Reactions (6.3) and (6.3a), but with N^{ϵ}-acetimidyl lysine at the carboxyl terminus of the carboxyl component.

Method

The N^{ϵ}-acetimidyl,O^{γ}-methyl-des-homoserine peptide (1 μmol), prepared by the first part of the Method of Example (5.3), was pre-treated by the Method of Example (6.5). It was dissolved in 2 μl of dimethylsulphoxide and 5 μl of redist. dimethylformamide (p. 218) added. To the solution was added 1 μl of a solution of 1-hydroxybenzotriazole (15.3 mg in 0.1 ml) in dimethylformamide and 1 μl of a solution of dicyclohexylcarbodiimide (20.6 mg in 0.1 ml) in dimethylformamide and the mixture was kept at 0 °C for 1 h. Meanwhile 0.5 μmol of the product of Example (6.2) was pre-treated by the Method of Example (6.4) and then dissolved in 10 μl of dimethylsulphoxide. The two solutions were mixed and the externally indicated pH adjusted to 7 with *N*-ethylmorpholine/dimethylsulphoxide (1 : 4, v/v); approximately 5 μl was required. The mixture was allowed to stand at 20 °C for 4 days. The reaction mixture was separated by the Sephadex column described in Method B of Example (3.7). The coupled product was checked by amino-acid analysis. Yield 30%, based on the amino component.

Notes

(1) The fragments are not very soluble in the solvents, except as their HCl salts, and it would be desirable to find others. In particular dimethylsulphoxide should be replaced, as its presence may lead to side-reactions. Couplings were attempted by mixing aqueous solutions of the two components, of *N*-hydroxysuccinimide, and of a water-soluble carbodiimide, but were not very successful (C. J. A. Wallace, unpublished results).

(2) The conditions were chosen by trial runs similar to those described in Stage 1 of Example (6.6).

(3) We find that 3,4-dihydro-3-hydroxy-4-oxo-1,2,3-benzotriazine (König and Geiger, 1972) is at least as effective as the 1-hydroxybenzotriazole. As with the latter reagent, it is necessary to carry out trial couplings to determine the optimum time for pre-activation of the carboxyl component.

Example (6.9). The semisynthesis of an analogue of cytochrome *c* by aminolysis of a carboxyl-terminal homoserine lactone

Source. Corradin and Harbury (1974a), Wallace (1976, 1978), and Wallace and Offord (1979).

Process involved. Reaction (6.1), in which the carboxyl component is a fragment derived from residues 1–65 and the amino component is a fragment derived from residues 66–104.

Method

Fragment 1–65 of cytochrome *c* was obtained by the freeze drying of the appropriate tubes from a second run of peak 4, Method B of Example (3.7), on the same Sephadex column. The fragment 66–104, obtained in Example (6.8) or directly from the protein (Note 4, Example (3.7)), was dissolved to a concentration of 10^{-4} mol/l in 0.1 M sodium phosphate, pH 7.0. Fragment 1–65 was then dissolved in this solution also to a concentration of 10^{-4} mol/l. The mixture was made up in a vessel chosen so as to be filled to capacity by the solution. Solid $Na_2S_2O_4$ (approximately 5 mg/ml) was then added, dissolved in the mixture, the vessel capped and allowed to stand at 20 °C for 24 h. The reaction mixture was then applied directly to the type of Sephadex column described in Example (3.7) (a column 2 cm × 135 cm, run at 40 ml/h is suitable for 10 ml of solution). The product is in peak 2 (p. 66).

Notes

(1) Corradin and Harbury (1974a), when coupling unprotected, unmodified fragments, found it necessary to ensure that the haem peptide was in the reduced form throughout. They obtained coupling yields of about 50%. Wallace (1976), using both unprotected and N^{ϵ}-acetimidyl-protected fragments, confirmed the desirability of reducing conditions and also obtained yields of 50% in 24 h. However, he obtained yields of about 12% in 6 days even without reduction.

(2) Corradin and Harbury (1974a) maintained the reduced state by the use of one equivalent of $Na_2S_2O_4$, in a rather complicated cell. The use of an excess $Na_2S_2O_4$, as described above, is simpler, but it is essential to avoid using too much, or the fragments precipitate. The rather small quantity of $Na_2S_2O_4$ does not provide much reserve reducing capacity, and it is essential to prevent the admission of O_2. It is best to choose a vessel of a volume only a little above that of the coupling mixture and to seal it tightly. The effects of even a small air space are seen after 24 h as a red (oxidized) ring at the top of the pink (reduced) solution. A light, water-immiscible liquid such as liquid paraffin can be used to fill the vessel up to the brim.

(3) It is possible to obtain fragment 66–104 in which Met^{80} has been converted to homoserine without cleavage. Wallace (1976) found that the efficiency

of coupling with this material was not improved by reduction of the haem peptide. In other words, the yield of coupled, 104-residue material was just over 10% after 6 days whether or not reducing conditions were employed.

(4) The fragment 1–65 was very largely in the lactone form, since it had only been in acidic solvents since its formation.

Example (6.10). The azide coupling of a CNBr fragment

Source. Offord (1972).

Process involved. Reaction (6.2). The carboxyl component (Lys-Gln-Thr-Ala-Ala-Ala-Lys-Phe-Gln-Arg-Gln-His-homoserine lactone) is derived from residues 1–13 of bovine ribonuclease.

Method

The peptide (10 μmol) was dissolved in trifluoroacetic acid (1 ml) and allowed to stand at 20 °C for 1 h. The solution was dried *in vacuo* and washed with methanol. It was then dissolved (4 mmol/l) in hydrazine (anhydrous)/dimethylformamide (1 : 9, v/v). The dimethylformamide was prepared as described on p. 218. After 6–10 min at 20 °C, the mixture was placed on the freeze drier. The dried material was redissolved in the minimum volume of dimethylformamide. A solution of dry HCl (10%, w/w) in dioxan (peroxide free) was added from a microcapillary until the externally indicated pH (p. 217) was 1. The mixture was cooled to −40 °C and 10 μmol of t-butyl nitrite was added as a 10% (v/v) solution in dimethylformamide. The mixture was left for 10 min and then, still at −40 °C, brought to an externally indicated pH of 7 with *N*-ethylmorpholine. Meanwhile 0.5 ml of a 0.2 M solution of tyrosine ethyl ester was prepared in dimethylformamide, and adjusted to an externally indicated pH of 7 with triethylamine. The tyrosine ethyl ester solution was cooled to −40 °C, the two solutions mixed, and allowed to warm up slowly to 4 °C. After 2 days at 4 °C, the product was passed down the gel-filtration column of Example (3.8). The first, peptide peak was subjected to paper electrophoresis and the band staining for both tyrosine and histidine (Note 6, Example (A.4)) eluted. Yield 10%.

Notes

(1) The initial treatment with trifluoroacetic acid is intended to convert any of the peptide that is in the open form to the lactone form. It can be omitted if the peptide is already substantially in the lactone form (cf. Note 4 to Example (6.9)).

(2) Anhydrous hydrazine can be obtained commercially or, with precautions against explosion, from hydrazine hydrate by the method of Kusama (1957).

The hydrazine stock should be kept in a number of small vessels. Each should be discarded after being used once in order to avoid the progressive contamination of the stock with water vapour. Such contamination must be avoided because of the danger of hydrolysis of the peptide and, more importantly, because of the difficulty of removing the resulting hydrazine hydrate. The use of the freeze drier in the Method above is sufficient to remove the hydrazine almost completely within a remarkably short interval, but the hydrate remains even after the dimethylformamide has gone.

(3) Wallace (1976) reported that some long peptides of cytochrome *c* were insoluble in dimethylformamide, but were brought into solution by the HCl/dioxan step.

(4) The HCl and t-butyl nitrite solutions can be diluted up to 10-fold with dimethylformamide in order to make the dispensing of small quantities easier.

(5) The azide coupling conditions are based on those of Visser *et al.* (1970).

(6) Offord (1972) and Wallace (1976) reported that the principal cause of the poor yields with this method is the tendency of the homoseryl azide to form the lactone, rather than to couple. The Curtius rearrangement (p. 5) which might have been expected, and which should have been detected easily by paper electrophoresis of the products of coupling between model substrates, was not observed.

(7) The conditions for the formation of the hydrazide were found by trial experiments on small quantities. Measurements of mobilities on paper electrophoresis were of the greatest assistance with the peptide described above (Offord, 1972) but it was not possible to use them to the same extent on longer peptides (Wallace, 1976).

The spray used by Offord (1972) for the hydrazide group (1 : 1 mixture of 0.018 M $FeCl_3$ and 0.015 M $K_3Fe(CN)_6$; Barton *et al.*, 1952) only gives unambiguous results when, as in the example above, the carboxyl component does not contain tyrosine. When this condition is met, the extent of hydrazinolysis can be confirmed by the ability of the appropriate paper-electrophoretic band to stain blue with the reagent. The formation of the azide abolishes the staining property.

(8) As stated in the text, the spontaneous aminolysis of the lactone is in general greatly superior as a coupling method to the one described here. However, it is not yet clear to what extent the spontaneous aminolysis has to be sterically assisted, and the azide method, possibly after some further development, might be found of use in instances in which steric assistance was lacking.

CHAPTER 7

Deprotection and purification

Deprotection

Even more than in Chapter 6, our options are limited by what has gone before. Table (7.1) indicates the conditions dictated by the various protecting groups. Our sole room for manœuvre is in arranging the details of the procedures so as to minimize the damage to the covalent and three-dimensional structure of the protein, and we must now examine the types of damage that can occur. There is little at this stage that distinguishes in principle a conventionally synthesized protein from a semisynthetic one, but we shall pay particular attention to those side-reactions that in practice are met with more in semisynthesis, because of the tendency to employ aqueous or semi-aqueous conditions as much as possible.

Hydrolytic damage at extremes of pH

Side-chain amides are the groups most labile to hydrolysis; the degree of sensitivity often seems to depend on the sequence in which they are found. Some workers believe that side-chain amides that are adjacent to polar residues, especially the basic ones, are the most labile. Also, the sequence -Asn-Gly- is notoriously labile both to amide loss and to the simultaneous transfer of the peptide bond from the α- to the β-COOH (cf. Reaction (1.20)).

$$\begin{array}{c} \text{CO{-}NH}_2 \\ | \\ \text{CH}_2 \\ | \\ \text{{-}NH{-}CH{-}CO{-}NH{-}CH}_2\text{{-}CO{-}} \end{array} \longrightarrow \begin{array}{c} \text{CO{-}NH{-}CH}_2\text{{-}CO{-}} \\ | \\ \text{CH}_2 \\ | \\ \text{{-}NH{-}CH{-}COOH} \end{array} \qquad (7.1)$$

Although peptide bonds are generally much more resistant to hydrolysis, certain sequences are cleaved under conditions of quite astonishing mildness (e.g. Hofmann, 1964). The cause is presumably the participation of groups adjacent in either the primary or the tertiary structure of the protein. The residues next in the primary structure to the site of cleavage are sometimes not those most often associated with residues to permit acid cleavage.

The danger of hydrolysis led long ago to the virtual abandonment of aqueous solutions of extreme pH as deprotecting agents for conventional synthesis. The use of NaOH solutions of almost any strength presents a considerable risk—this has not prevented quite severe NaOH treatments being used, with un-

Group	Reagent						
	HF	trifluoro-acetic acid*	Trifluoro-acetic acid	1 M NH_4OH, or triethylamine carbonate (pH 10.6)	NH_3/acetic acid (pH 11.3) (Example (7.5))	Dilute NaOH (pH > 11.5)	Other reagents for removal
N-Acetimidyl-	?	Stable	Stable	?	Removed	Stable	Hydrazine
N-Trifluoroacetyl-	Stable	Stable	Stable	Removed	Removed	Removed	Hydrazine, aqueous piperidine
N-Phthaloyl-	Stable	Stable	Stable	Stable†	Stable†	Removed	Hydrazine
N-Cbz- and N^{im}-Cbz—NH—CH(CF_3)-	Removed	Removed	Slowly removed	Stable	Stable	Stable	Reduction
N-Boc- and N^{im}-Boc—NH—CH(CF_3)-	Removed	Removed	Removed	Stable	Stable	Stable	§
N-Maleyl-	Stable	Stable	Stable	Stable	Stable	Stable	Dilute aqueous acid
—S—SO_3^-	Stable	Stable	Stable	Stable	?	Removed	Thiols, Hg^{2+}
Methyl ester	Stable	Stable	Stable	Removed	Removed‡	Removed	
t-Butyl and t-amyl esters	Removed	Removed	Removed	Stable	Stable†	Stable	§
Benzyl ester	Removed	Removed	Stable	Slowly removed	Removed‡	Removed	Reduction
p-Methoxybenzyl ester	Removed	Removed	Removed	Slowly removed†	Removed†‡	Removed	Reduction
4-Methoxy-3-sulphobenzyl ester	Removed†	Removed†	Removed	Removed	Removed†‡	Removed	
Diphenylmethyl ester	Removed	Removed	Removed	Slowly removed†	Removed†‡	Removed	Reduction

* Preferred to HBr/acetic acid because of the danger of acylation of aliphatic —OH groups.
† Predicted result.
‡ Would leave at least some amide.
§ Stable to hydrazine and to reduction.

The *N*-methylsulphonylethyloxycarbonyl group is stable to trifluoroacetic acid. It is removed in aqueous alkali and in anhydrous solutions of NH_3, triethylamine, and piperidine.

known consequences, in certain semisyntheses (see Chapter 8). The more labile amide groups are at risk even with the other basic media in Table (7.1). Indeed, some amide groups are so labile that a protein, even when isolated from living material by the most gentle methods, is often found in more than one form, differing solely in their amide content. Insulin and cytochrome *c* are well-known examples in which small quantities of des-amide forms are normally present. The biological activity of these variant forms is often unchanged and so amide loss, while always undesirable, need not be a total disaster. However, these observations (which imply equally that conventionally synthesized materials must be subject to similar suspicions) mean that we must be sure that we know if amide loss has occurred and if so at what sites, and to what extent. The methods for checking the state of the amide group are described in Appendix 1.

The acidic media in Table (7.1) are unlikely to cause hydrolysis if they are strictly anhydrous.

Amidation

Basic aqueous solutions that contain NH_3 in concentrations of 1 mol/l and above should not be used until the ester groups have been removed. If this precaution is neglected, there will be some conversion of the side-chain esters of glutamic and aspartic acids to glutamine and asparagine. A terminal ester, similarly, can be amidated. The conversion occurs most readily with methyl esters but is noticeable with others, such as the benzyl esters (R. E. Offord, unpublished work). Tertiary esters are presumably more or less immune.

Transpeptidation

We saw in Chapter 1 that the cleavage of methyl-ester groups by OH^- can lead to $\alpha \to \beta$ or $\alpha \to \gamma$ transpeptidation (Scheme (1.20)). Bruckner *et al.* (1959) report that this damaging side-reaction is suppressed if the deprotection takes place in the presence of Cu^{2+}. Transpeptidation may be less serious with the deprotection of large molecules than it clearly is with the small model peptides studied by Kovács *et al.* (1955). C. J. A. Wallace, D. E. Harris and R. E. Offord (unpublished) found no evidence for its having occurred in the deprotection of the methyl ester of cytochrome *c* at pH 10.6. Nonetheless transpeptidation should be a cause for concern. As with amide loss, methods exist, drawn from sequence analysis, by which it can be detected (Appendix 1).

The N→O acyl shift

Strong acids can promote the transfer of a peptide-bond —COOH group from the α-NH_2 of serine or threonine to the side-chain —OH. This N→O acyl shift is readily reversed in aqueous solution at pH values only a little above 7.

$$\begin{array}{c}R_1 \qquad\qquad\qquad CH_2OH \\ -NH-CH-CO-NH-CH-CO- \underset{base}{\overset{acid}{\rightleftharpoons}}\end{array}$$

$$\begin{array}{l}\qquad R_1 \\ -NH-CH-CO-O-CH_2 \\ \qquad\qquad\qquad H_2N-CH-CO-\end{array} \quad (7.2)$$

Since the shift is reversible and there is little chance of racemization, it does not really constitute a serious problem. Nonetheless it occurs to a marked extent in at least one deprotection scheme, which as a consequence includes provision for the reversal of the shift (see Example (7.7) below).

Acid-mediated damage to side-chains

The acids in Table (7.1) can, as we saw in Chapter 1, damage some sensitive residues, particularly tryptophan. The damage can be minimized by using only very pure reagents. If diazoalkanes were used at an earlier stage, the degree of acid-induced damage of tryptophan will be much increased unless the last traces of the by-products of the esterification are removed (Offord *et al.*, 1976). Scavengers (p. 19) help protect the sensitive residues, but their use can pose a threat to the non-covalently maintained structure of proteins (p. 146). Even if they are used some damage to tryptophan is inevitable. Rees and Offord (1976a) examined ways, including the use of scavengers, for minimizing deprotection damage to tryptophan after a semisynthesis.

Damage involving cysteine and cystine

Some cysteine derivatives undergo β-elimination in alkaline media: for example, under strongly basic conditions

$$\begin{array}{l}S-CH_2-C_6H_5 \\ CH_2 \\ -NH-CH-CO-\end{array} \rightleftharpoons -NH-\overset{\overset{CH_2}{\|}}{C}-CO- + HS-CH_2-C_6H_5 \quad (7.3)$$

These derivatives must be avoided, or cleaved in advance of any exposure to strong base. Even the unprotected —SH group could be damaged at very high pH.

The problem of β-elimination is minor in comparison with those posed by the tendency of free cysteine residues to be oxidized spontaneously to form disulphide bridges.

$$\begin{array}{c}SH \\ SH\end{array} \xrightarrow{[O]} \begin{array}{c}S \\ | \\ S\end{array} \quad (7.4)$$

If the protein in question does not normally possess disulphide bridges, this reaction must be prevented until the folding up of the protein removes the thiol groups out of reach of each other. Oxygen and catalysts of oxidation must be rigorously avoided, or their effects must be overcome by the addition of a large excess of a thiol of low molecular weight. It is essential to remove and exclude O_2 thoroughly: a 1% solution of a small protein might only require 20 μl of O_2 per ml for complete oxidation.

If the protein does have disulphide bridges in the native state, we have to ensure that each cysteine residue finds its correct partner. We must prevent the formation of spurious intra- and intermolecular disulphide bridges both while the bridges are first forming or later, by disulphide interchange.

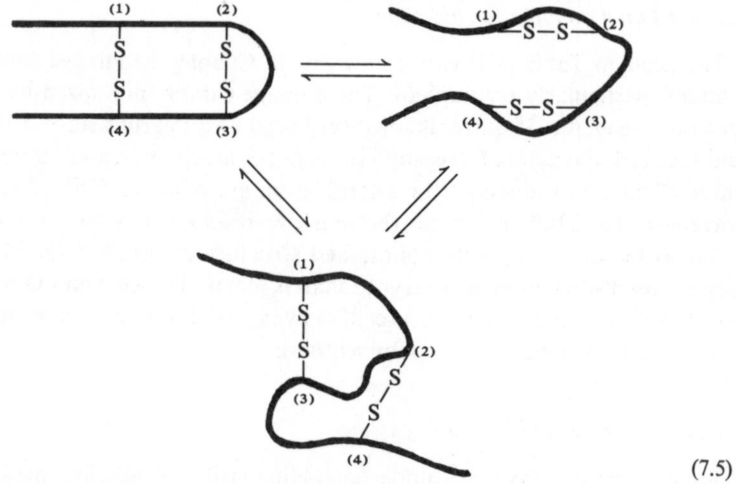

(7.5)

The formation of incorrect disulphide bridges within the molecule will lead to an incorrect tertiary structure and the protein will almost certainly be inactive. Intermolecular disulphide bridges will in addition give rise to an insoluble polymer of high molecular weight.

Fortunately, the formation of the correct bridges is usually assisted sterically by the non-covalent forces that produce and then stabilize the tertiary structure of the protein. The same forces tend to lead to the correct bridges being retained once formed. If we are dealing with such a protein, we have only to keep the bridges reduced until denaturants are removed and (to avoid intermolecular bridging) we must keep the concentration of protein low. Conventional syntheses and semisyntheses that involve cleavage of disulphide bridges have always relied for this step on conditions selected by prior experiments on the reduction and reoxidation of disulphide bridges in the natural protein. The efficiency of the process can then be made quite high. However protein analogues, which necessarily have a different balance of non-covalent forces than in the natural molecules, may not tend so strongly to fold up in the correct way. The position is worse still if the protein is derived *in vivo* from a precursor by proteolytic cleavage. The molecule that we synthesize may not possess all the features that originally brought the correct residues together in the precursor. This last problem is very severe in work on insulin, and the efficiency of reoxidation of the natural chains is poor and that of analogues, particularly conventional, but also semisynthetic, is worse. The option, in many semisyntheses, of keeping the original disulphide bridges intact throughout is one of the greatest potential advantages of the whole approach.

Non-covalent damage: denaturation

In the chemical synthesis of a small molecule, one is most concerned to form the correct covalent bonds and, once the product has been made, to guard against covalent decomposition or damage. It is sometimes difficult to realize that in work on proteins we add to these worries others of equal importance about the *non-covalent* structure. Proteins owe most of their extraordinary properties to their being able to combine and use non-covalent interactions on a scale that would be impossible with smaller molecules. Therefore, even if all the peptide bonds and disulphide bridges are correct we must try to ensure that the three-dimensional, non-covalently stabilized structure is the correct one: in other words our product must not be denatured.

When one considers how profoundly most deprotecting agents differ in their solvent characteristics from those of physiological media, it is surprising that protein emerges from deprotection able to fold up correctly and to function. In fact many proteins can be recovered in a functioning state from an aqueous solution of a range of pH of 1–11 or wider; from liquid HF; from trifluoroacetic acid and from many denaturing organic solvents. They presumably lose their correct structure and recover it when the conditions become more physiological. Other proteins may be less hardy. These can be trapped, kinetically or thermodynamically, in a wrong configuration, either by non-covalent forces or by disulphide interchange. It is therefore highly desirable, before embarking on any sort of synthesis, to test the response of samples of the natural protein to all the conditions that will be encountered in the proposed scheme of reactions.

The $\Delta H^{\ddagger}$ of protein denaturation is such that the rate of the process increases very much more rapidly with temperature than does that of practically any other chemical or physical change. Other conditions, such as pH, or the concentration of some reagent, may similarly produce a threshold of violent change in the tertiary structure. Since there is always the danger that denaturation, once it has occurred, will not readily be reversed it is probably best not to attempt to force the rate of deprotection by intensifying the conditions beyond what is normal.

As a counterweight to all these warnings it should not be forgotten that, apart from the well-known examples of stability already mentioned, many proteins are exposed *in vivo* for long periods of time to conditions somewhat removed from those that we are accustomed to call physiological. The proteins of microorganisms from hot springs thrive at temperatures above 70 °C. Even mammalian proteins come from an environment considerably warmer than that of the laboratory, let alone the cold room. Extremes of pH are not unknown *in vivo*: the pH of the stomach is about 2, whilst that of egg white is 9.5. Proteins function happily in both these environments for extended periods under warm conditions.

Deprotection in practice

Example (7.1) gives a general method for deprotection by trifluoroacetic acid. It is notable that a scavenger added to protect against covalent damage can enhance the degree of denaturation. Example (7.2) gives conditions, which are probably among the mildest possible, that are particularly applicable when very small quantities are involved.

Examples (7.3) and (7.4) give generally useful conditions for the removal of base-labile groups. Some of the traditional reagents for the removal of such groups as the methyl and trifluoroacetyl are probably too strongly basic for most proteins.

Example (7.5) gives conditions for the removal of the acetimidyl group from the ε-NH_2 group of lysine. These are rather vigorous conditions and, since the ε-acetimidyl lysine residue is likely to be very similar in its contribution to the structure and function of the protein to that of lysine, many workers are content to leave it in position (see Chapter 8).

Example (7.6) gives conditions for the removal of the phthaloyl group, Example (7.7) for the removal of acyl methionyl groups, Example (7.8) gives conditions for the deprotection of maleyl groups and Example (7.9) for the deprotection of cysteine sulphonic acid.

There are very many acid-labile groups, but there is a great need for more —NH_2 and —COOH protecting groups that are stable to acid (so as to be compatible both with acid-labile groups and with the use of the Edman reaction) but are labile to moderately basic conditions, in the region of pH 9–10. Acid-stable groups that are labile to mild reagents of other kinds would also be extremely useful.

Purification of the product

The final sample, synthetic or semisynthetic, is almost bound to contain some denatured material. There will also be some inadequately deprotected, or otherwise covalently incorrect, molecules. How can the wanted product be obtained from this mixture in the maximum purity consistent with there being a reasonable yield?

All the normal methods of preparative biochemistry are available, and separations based on size (in particular gel filtration) and charge are likely to recommend themselves. The literature cited in Chapter 8 reveals how the problem, as it arises with semisynthetic products, has been met (or avoided) in practice.

Enriching for native protein

It is best to include in the purification scheme for the product a system that will reject molecules that, while possibly covalently correct, do not have the proper tertiary structure. Crystallization, though it is rightly used as little as possible in the general purification of macromolecules, can provide a means of enriching the sample in native material. Example (7.10) describes a useful method for crystallizing insulin which is also applicable to some, but not all, of its modified derivatives.

The native form of the protein can be removed from denatured material by specific antibodies. Only molecules that have the correct conformation of the major immunological determinants will be bound. Example (7.11) illustrates the immunological purification of a semisynthetic insulin on a column of immobilized anti-insulin. Affinity chromatography (see the Review section of the Bibliography) could be used in the same way. This technique depends on immobilizing on a column a molecule (such as, for an enzyme, a substrate analogue), for which the native form of the protein has a high and preferably specific affinity. As with the immunoabsorbant column, the native material is retained and can usually be eluted from the column under mild conditions.

EXAMPLES

Example (7.1). Deprotection with anhydrous trifluoroacetic acid

Source. A standard method.

Process involved. Reactions of the type of (1.13). Table (7.1) indicates the groups that are removed.

Method A

The protected material was dissolved in the acid to a concentration of about 1% (w/v) and allowed to stand at 4 °C, usually for 40 min to 1 h. (Alternatively (e.g. Harris, 1977) at 20 °C for 30 min.) After this period the solution was taken to dryness *in vacuo*. Traces of acid can be removed by dissolving or suspending in methanol, and taking to dryness again.

Method B

The procedure was carried out as described in Method A except that the deprotected material was recovered from the acid solution by precipitation with peroxide-free ether.

Notes

(1) Scavengers (p. 19) can be added to the trifluoroacetic acid. A 50-fold molar excess of anisole over the groups to be protected is usual. The anisole may enhance the denaturation of some proteins and it may be desirable to restrict its concentration to 1–2% (v/v). This in its turn may necessitate the use of a larger volume of acid. If denaturation is a danger, then Method B must be used when anisole is present, as it is less volatile than the acid and will increase in concentration on drying down.

(2) Rees and Offord (1976a) investigated various acidic regimes, with and without scavengers, for the deprotection of tryptophan-containing peptides. They decided that there was little advantage in any more complicated procedures than those in the Method above and that the use of scavengers did not materially improve the recovery of tryptophan.

(3) Insulin derivatives do not appear to be harmed if the treatment takes place at 20 °C.

(4) A few minutes' exposure is generally sufficient to remove the Boc-, t-butyl, and 1,1-dimethylpropyl groups, and (from the side-chains of glutamic-acid residues) the *p*-methoxybenzyl group. The *p*-methoxybenzyl group is more slowly removed from the side-chains of aspartic-acid residues (2–3 h exposure at 4 °C is required) and more slowly still from the —COOH of *N*-maleyl groups (Rees and Offord, 1976a).

(5) Trifluoroacetic acid has a low heat of evaporation and is usually removed with remarkable rapidity *in vacuo*. However, for some reason a small fraction will sometimes remain and defy all attempts to evaporate it. The result is a strongly acidic brown oil of proven ability to denature. This phenomenon was observed by Halban and Offord (1975) even when highly purified acid was used.

(6) The same precautions are necessary when collecting and washing the precipitate in Method B as were described in Note 2 to Example (3.1).

(7) The denaturing properties of anisole would be even more serious if HF were to be used rather than trifluoroacetic acid. The physical danger of handling HF makes it obligatory to remove the acid by evaporation and the consequent increase in the concentration of anisole could not be avoided.

Example (7.2). Deprotection with anhydrous trifluoroacetic acid on a very small scale

Source. Halban and Offord (1975), and unpublished work.

Process involved. As in Example (7.1) except that the acid solution is not dried down, but diluted directly into a buffered solution of carrier protein.

Method

The product of Example (6.3) was dissolved (6 μg/μl) in anhydrous trifluoroacetic acid containing 1% (w/v) serum albumin and allowed to stand at 20 °C for 45 min. The solution was then added to 100 vol. of 50 mM $(NH_4)_2HPO_4$ (adjusted to pH 10 with 1 M NH_3 and containing 5% (w/v) serum albumin). The pH fell to near neutrality and was adjusted to pH 7.3 with 1 M NH_3 or 1 M H_3PO_4.

The buffered solution was then subjected to gel filtration in 5 mM Na-phosphate buffer (pH 7.3) on a column of Sephadex G-25 (0.8 cm × 66 cm). The protein, which eluted in the void volume, was cleanly separated from the remaining traces of by-products of the active ester. The concentration of the [^{3}H]insulin in the pooled peak tubes was approximately 1 μg/ml (about 3×10^6 c.p.m./ml at high specific activity). The serum albumin concentration was checked spectrophoretically and adjusted if necessary to 0.5% (w/v). The product was checked by isoelectric focussing, Example (A.8), and by the depression of blood-sugar levels in mice. The material behaved indistinguishably from native insulin in both respects.

Notes

(1) In contrast with analogous larger-scale preparations of non-radioactive material, the drying down of a trifluoroacetic acid solution of the ^{3}H-labelled product leads to loss of side-chain amide groups. Bands corresponding to [^{3}H]insulin lacking one, two and three amide groups are seen on isoelectric focussing. The Method above prevents this damaging side-reaction.

(2) The serum albumin (which should be immunologically compatible with the species to which the ^{3}H is to be administered) must be of the best available grade. Some samples contain enzymes that degrade insulin. Those supplied by Miles (Pentex) Schering and Armor have been found satisfactory. If human albumin is used, it should be certified free of Australia antigen by a reputable supplier.

(3) The reason for the use of the ammonium salt in the buffer is to accelerate the breakdown of any residual active ester. An ethanolamine buffer would probably be even more effective.

(4) In the original preparation (Halban and Offord, 1975), the buffer into which the trifluoroacetic acid solution was diluted had been previously adjusted to pH 7.3. The pH fell to 2 on addition of the acid solution and had to be restored to neutrality. The adjustment in the Method, above, is much smaller in magnitude.

Example (7.3). Removal of base-labile protecting groups: (A) removal of the *N*-trifluoroacetyl group by 1 M NH_4OH; (B) removal of the methylsulphonylethyloxycarbonyl group by NaOH

Source. (A) Borrás (1972). (B) Tesser (1975), and Geiger *et al.* (1975a).

Process involved. (A) Reaction (1.16). The NH_4OH is simply a source of OH^-. (B):

$$CH_3{-}SO_2{-}CH_2{-}CH_2{-}O{-}CO{-}NH{-}R \xrightarrow{\text{base}}$$
$$CH_3{-}SO_2{-}CH_2{-}CH_2OH + CO_2 + H_2N{-}R \qquad (7.6)$$

Method

A. The sample was dissolved in 1 M NH_4OH (approximately 1 μmol of trifluoroacetylamino groups/ml). The pH was checked externally with pH paper. The solution was dried *in vacuo* after 24 h at 20 °C. Insulin derivatives were characterized by isoelectric focussing and crystallization (Examples (A.8) and (7.10)).

B. The sample was suspended (approximately 50 mg/ml) in 1.7 ml dioxan/water (1 : 1, v/v) and brought into solution by the addition of 0.25 ml of 2 M NaOH at 0 °C. After about 30 s the solution was brought to pH 3 with glacial acetic acid and purified by gel filtration.

Notes

(1) Goldberger and Anfinsen (1962) used 1 M piperidine at 0 °C for the removal of the trifluoroacetyl group, but the pH (above 12) is too high for most proteins. This is also true of the 0.2 M NaOH used by Weygand and Csendes (1952) and the 0.2 M $Ba(OH)_2$ used by Zahn and Pätzold (1963). Fanger and Harbury (1965) used mild conditions for the deprotection of trifluoroacetylated cytochrome *c*: 0.15 M K_2CO_3, pH 10.7 for 30 h at 25 °C.

(2) Borrás (1972) found that 0.1 M NH_4OH was just as effective in Method A if the treatment was prolonged to 40 h at 20 °C.

(3) Trifluoroacetyl derivatives of insulin sometimes fail to give crystalline material after deprotection (Borrás, 1972). The reason is not known, but the phenomenon need not necessarily prevent the use of the method with other molecules.

(4) Esters (except tertiary ones) should be removed before the use of Method A treatment because of the danger of partial amidation. Methyl esters are particularly susceptible to this side-reaction.

(5) The conditions of Method B correspond to a pH of about 13. Geiger *et al.* (1975a) describe another procedure, in which the product (an insulin derivative) was suspended in water/dioxan/methanol (2 : 1 : 1, by vol.) to a concentration of about 50 mg/ml. To 2 ml of this solution 2 M NaOH was added (approximately 0.15 ml) to produce a pH of about 10. Deprotection was over in 20 min at 0 °C.

(6) Van Nispen *et al.* (1977), working with an ACTH derivative, used dioxan/methanol/water (30 : 9 : 1, by vol.) 0.2 M in NaOH for 15 s.

(7) The control of the neutralization step in Method B calls for considerable care. It has to be done quickly, preferably without overshooting, and yet the excessive production of heat must be avoided.

(8) Scheme (7.6) is tentative. See Tesser (1975).

Example (7.4). The deprotection of methyl esters by saponification

Source. C. J. A. Wallace, unpublished experiments and Wallace and Offord (1979).

Process involved.

$$R_1{-}CO{-}O{-}R_2 \xrightarrow{\text{base}} R_1{-}COOH + R_2OH \qquad (7.7)$$

Method

An aqueous solution of redist. triethylamine (2%, w/v) was brought to pH 10.5 with solid CO_2. The material to be deprotected was suspended or dissolved in the solution (10 μmol/ml) and, after a check on the pH, the mixture allowed to stand for 24 h at 20 °C. The solution was then neutralized with 1 M acetic acid and freeze dried.

Notes

(1) Over-enthusiastic addition of CO_2, apart from depressing the pH, can lead to the freezing of the liquid in the pH electrode. The envelope shatters.

(2) The above Method is the best for saponifying esters of fragments of cytochrome *c* but gives an insoluble product when applied to esters of the complete cytochrome. By contrast, the ester of the cytochrome regains the majority of its biological activity after saponification at pH 10.5 under pH control by the addition of NaOH to an aqueous solution of the derivative in an autotitrator. It is therefore important to carry out trial experiments when saponifying a product for the first time.

(3) Gattner *et al.* (1978), working with the methyl ester of insulin, found it best to use 0.1 M NaOH at 10 °C for 45 min. Weitzel *et al.* (1978) used pH 11.8, 4 °C.

Example (7.5). The removal of the acetimidyl group

Source. Ludwig and Byrne (1962).

Process involved.

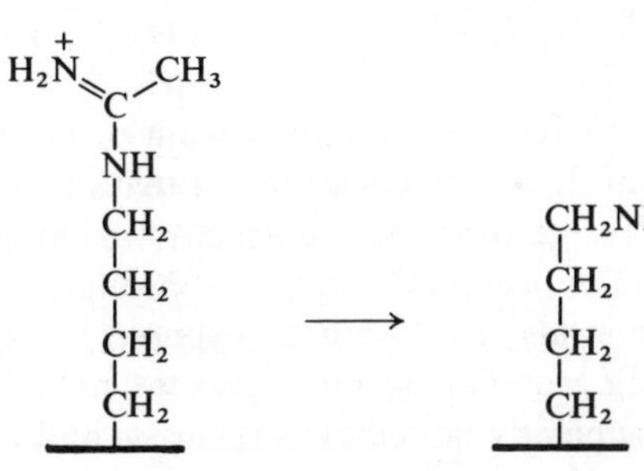

$$\underset{\underline{\quad}}{\overset{H_2\overset{+}{N}{=}C(CH_3){-}}{NH{-}CH_2{-}CH_2{-}CH_2{-}CH_2{-}}} \longrightarrow \underset{\underline{\quad}}{NH_2CH_2{-}CH_2{-}CH_2{-}CH_2{-}} \qquad (7.8)$$

Method

The material to be deprotected was dissolved (0.4–10 μmol/ml) in a reagent prepared by the addition of glacial acetic acid to concentrated NH_3 solution (35%, w/v) until the apparent pH (glass electrode) was 11.3 (approximately 1 ml of acetic acid to 15 ml of NH_3 solution). The solution was left for 16–24 h at 20 °C, diluted by the addition of 10 ml water, and freeze dried.

Notes

(1) It is essential to remove esters before this treatment (cf. Note 4 to Example (7.3)).

(2) If appropriate, the sample can be dissolved first in glacial acetic acid and the concentrated NH_3 added (with cooling) until the correct pH is reached. Because of solubility problems (cf. p. 218) this approach was found to be essential when deprotecting cytochrome *c* derivatives (Wallace, 1976).

(3) Ludwig and Hunter (1967) suggested other methods for the nucleophilic displacement of the N^{ϵ}-acetimidyl group but observed side-reactions during the deprotection of acetimidylated insulin under conditions of very high pH or of elevated temperature.

Example (7.6). The removal of the phthaloyl group

Source. Geiger and Langner (1972, 1975); R. Geiger, private communication.

Process involved. Reaction (1.18).

Method

The material to be deprotected was dissolved in phenol/water (4 : 1, w/v) at a concentration of 8 μmol of phthaloyl groups/ml. Hydrazine hydrate was then added (20 μl for every 1 ml of the phenol mixture). After 16 h at 40 °C, the mixture was precipitated by the addition of ether, and purified by one of the Methods of Example (2.9).

Notes

(1) The phenol protects insulin derivatives from denaturation in a manner which is as yet unexplained. Borrás (1972) found that in the absence of phenol a concentration of hydrazine (even as the acetate, pH 8) sufficient to remove phthaloyl groups broke the disulphide bridges of insulin.

(2) Phthaloylated materials should not be exposed to aqueous solutions of base, since the protecting group opens up to give the phthalamidic-acid derivative, which cannot subsequently be removed (Hanson and Illhardt, 1954).

(3) Pechère and Bertrand (1977) report that, in common with the maleyl group and many of its analogues (Table 2.1), the phthaloyl group can be removed by exposure (in solution or in suspension) to an aqueous solution buffered to pH 3.5. Pechère and Bertrand added, with vigorous stirring, 0.2 ml of 1 M NaOH to each ml of solution (maximum protein concentration 4 mg/ml) followed by 0.56 ml of 1 M formic acid. The mixture was then incubated at 50 °C for 48 h, with constant gentle stirring if there was insoluble material present. Compare Example (7.8).

Example (7.7). Removal of acyl methionyl protection by CNBr

Source. Saunders and Offord (1977a, b).

Process involved. Reaction (2.6).

Method

The insulin derivative (usually a diacyl methionyl, mono-Boc- derivative) was dissolved (5 mg/ml) in aqueous formic acid (70%, w/v). Cyanogen bromide (5 mg/mg insulin) was added and the reaction allowed to proceed at 20 °C for 20 h. The mixture was freeze dried after dilution with water.

Notes

(1) Notes 2 and 3 to Example (3.6) apply.

(2) A check on the deprotected material by the usual analytical tests usually indicates that deprotection is substantially complete. However, the prolonged formic-acid treatment appears to produce either a N→O acyl shift (p. 138) or formylation of the —OH groups of serine or theronine. A deprotected insulin derivative will not crystallize until the change has been reversed by treatment with an aqueous solution of pH between 8 and 9 for 1–2 h at 20 °C (see, for example, Note 7 to Example (7.10)).

(3) Saunders and Offord (1977a) report that the use of a simple acyl group to protect the methionine, such as benzoyl, gives less satisfactory results than urethan protection. Methionine residues protected by the acyl groups are much more likely to undergo conversion to the homoserine derivative without cleavage of the peptide bond.

Example (7.8). Removal of maleyl groups

Source. Rees and Offord (1976a) based on Butler *et al.* (1969).

Process involved. Intramolecularly catalysed hydrolysis.

Method

The protected material was dissolved (10 mg/ml) in pyridine/acetic acid/water (1 : 10 : 89 by vol., pH 3.5). The reagent was made 8 M in urea if required to solubilize the material. The solution was allowed to stand for 100 h at 37 °C. The product is recovered by dialysis or, if the solubility properties permit, by centrifugation after dilution into 25 vol. of distilled water.

Notes

(1) Butler *et al.* do not consider it necessary to bring insoluble materials into solution.

(2) The long exposure to urea does not carry much risk of cyanation, because the pH is low.

Example (7.9). The removal of the *S*-sulphityl group

Source. A standard technique.

Process involved. Reaction (2.13).

Method

The material was dissolved in an appropriate medium (for example an aqueous buffer at pH 8) 0.2 M in 2-mercaptoethanol. After 2 h at 37 °C the product was recovered from the solution by dialysis or some other appropriate method.

Notes

(1) The method for the recovery of the deprotected material depends on the purpose to which it is subsequently to be put. If the —SH groups are to be allowed to form disulphide bridges, the reduced solution should be adjusted to the appropriate concentration and the oxidation promoted as the dialysis proceeds. If the —SH groups are to be alkylated (e.g. for analytical purposes), the best course is to bring the concentration of 2-mercaptoethanol down by dialysis, but not reduce it to zero, in order to prevent oxidation. An appropriate excess of alkylating agent should then be added, sufficient both for the protein —SH groups and the remaining mercaptoethanol.

(2) Dithiothreitol (Cleland, 1964) is an efficient and less unsavoury substitute for 2-mercaptoethanol.

Example (7.10). The crystallization of a semisynthetic insulin

Source. Schlichtkrull (1956, 1958).

Process involved. Formation of rhombohedral crystals of the hexameric form of insulin in which two atoms of Zn are bound per hexamer.

Method

The deprotected derivative (10 mg) was dissolved in 0.02 M HCl (1 ml), to which was added, in order, 0.12 M $ZnSO_4$ (0.1 ml), 0.2 M trisodium citrate (0.5 ml), acetone (0.3 ml), and water (0.1 ml). The solution was adjusted to between pH 6.1 and 6.3 with 0.1 M HCl. The vessel was stoppered and left at 20 °C. Small crystals of insulin were deposited in 1–72 h.

Notes

(1) The acetone should redissolve the precipitate formed on the addition of the previous two solutions. Provided the correct amount of insulin was taken, any significant quantity of precipitate remaining after the addition of the acetone is likely to derive from denatured material. The precipitate can be removed either before or after the adjustment of the pH. It is probably better to do so before the adjustment as very occasionally crystals form so rapidly after the adjustment that some might be removed with the precipitate. It must be remembered, of course, that a semisynthetic analogue might be intrinsically less soluble.

If only small quantities of precipitate remain after the addition of the acetone and the water, it can sometimes be clarified by the addition of a few drops of 0.75 M NaOH before the adjustment to pH 6.1–6.3.

(2) The solubility of the insulin falls considerably with a drop in temperature. This effect can be exploited to promote crystallization (although there is always the danger that too rapid a deposition from solution will lead to the formation of amorphous material), for example if the operation described in Note 1 has depleted the quantity of insulin in solution.

(3) If it is desired to grow large crystals, the mixture can be taken to 50–60 °C and allowed to cool slowly in an insulated container (Adams *et al.*, 1969).

(4) If the acetone is replaced by 0.3 ml of an aqueous solution of phenol (6.7%, w/v) the insulin crystallizes, rather less well, in the monoclinic form. If it is replaced by NaCl solution (7%, w/v) it crystallizes with four Zn^{2+} ions per hexamer. The packing of the molecules differs from one crystal form to another and an analogue that is unable for steric reasons to crystallize in one form may do so in another.

(5) If the material refuses to crystallize, it is worthwhile trying pH values a little outside the range 6.1–6.3.

(6) Crystals can be separated from amorphous material by sucrose-gradient sedimentation.

(7) When the material comes from deprotection by the Method of Example (7.7), the crystallization mixture should be adjusted to pH 8.5 with 0.2 M NaOH and allowed to stand for 2 h at 20 °C before the adjustment to pH 6.1–6.3.

Example (7.11). Purification of a semisynthetic product by immunoadsorption

Source. Halban *et al.* (1976), scaled down from the method of Akanuma *et al.* (1970).

Process involved. (a) The covalent attachment of an antibody to beaded agarose. In the present instance the material used was unfractionated serum of guinea pigs that had been immunized against insulin. (b) The immobilized antiserum will, when packed in a column, only retain those molecules with the appropriate immunological activity. These molecules can subsequently be eluted by, for instance, a change in pH. The sample in the present instance was semisynthetic [^{3}H]insulin.

Method

Cyanogen bromide activated agarose beads (CNBr-activated Sepharose 4B, Pharmacia, Ltd.) were washed by suction on a cellulose acetate filter with 1 mM HCl and then, quickly, with cold 0.14 M NaCl/0.1 M $NaHCO_3$. A quantity equivalent to 30 mg dry weight (approximately 0.1 ml swollen volume) was suspended in 0.1 ml of the same buffer to which had been added 0.01 ml of the serum. After 20 h at 2 °C, the beads were washed with 0.14 M NaCl and then with 0.01 M Tris-HCl, pH 8.6. They were then packed into a small column (0.3 cm × 1.5 cm) and equilibrated with 0.01 M Tris-HCl (pH 8.2–8.4). All subsequent operations were carried out at 20 °C. The sample ([^{3}H]insulin prepared as in Example (7.2), 2.5 ng) was allowed to run into the column. The column was eluted with the Tris–albumin buffer under a 5–10 cm head of pressure. The radioactive material that emerged at once was discarded. The bound material was eluted with 1 M acetic acid. The radioactivity emerged as a peak at about the time that the pH of the effluent was falling. The eluted material was dialysed against distilled water. The specific immunoreactivity of this fraction was 50 times that of the material that had passed through the column unbound.

Notes

(1) It is, clearly, essential to avoid amine-containing buffers during the coupling step, since such molecules would compete with the proteins' —NH_2 groups for attachment. It is possible, but not essential, to ensure that all potential sites of attachment have reacted by exposing the product of part (a) of the method to 1 ml of 1 M ethanolamine-HCl, pH 9, for 2 h at room temperature. In the present instance the Tris buffer will have the same effect.

(2) Scatchard plots showed that the immunoadsorbent, as prepared above, had a capacity of 50 ng insulin/ml of swollen gel. The effective binding constant was approximately 10^{-7} M. No doubt the capacity, and possibly the binding constant, could be substantially improved by using a purified gamma-globulin

fraction of the antiserum. Also, it is quite often advantageous to attach the protein to the solid support by means of a side-chain of approximately 10 Å in length. Agarose derivatives with such side-chains, ready-activated for protein coupling, are commercially available. (For further information on the preparation of immobilized adsorbents see Jakoby and Wilchek (1974) and Mosbach (1976).) It is also a simple matter to scale up the present Method.

(3) It is very convenient to use the pre-activated agarose beads, but the activation (by the method of Axén *et al.*, 1967) is quite easy to perform, with a little practice.

(4) The Manufacturers' instructions suggest that the protein should be coupled to the adsorbent in 0.1 M $NaHCO_3$/0.5 M NaCl. The composition of this buffer is probably not critical so long as it is mildly alkaline, free of amines, and of reasonable ionic strength. By contrast it is most important to use the 1% (w/v) albumin buffer in the immunoadsorption and elution steps: failure to do so can lead to non-specific adsorption of protein.

(5) [^{3}H]Insulin was chosen to illustrate this technique because it had been mentioned so often before. In practice, insulin can be purified by a less general, but effective, method. The sample is chromatographed on columns of Sephadex G-50 in 0.2 M glycine (sodium)/serum albumin, 0.25% (w/v), pH 8.8. Denatured material emerges first, possibly because it binds more tightly to the albumin than does the native hormone.

CHAPTER 8

A review of published work on semisynthesis

Any derivative of a natural protein that is formed by the use of a synthetic reagent—an acetylated trypsin, for example—could, taking the term literally, be called semisynthetic. In practice the expression has a more restricted meaning. Smyth (1975), for example, used it* in the sense of a synthesis that utilizes a *fragment* of a natural polypeptide as a prefabricated unit in the construction of a larger one. I wish to follow this definition, but if taken too literally it not only excludes the very simple protein derivatives like the acetyl trypsins, but also those in which a specific transformation has been effected on a single side-chain. Examples of such work are the conversion of the amino-terminal serine residue of the hormone ACTH into glycine (Dixon and Weitkamp,1962), the conversion of the —CH_2OH of the active-site serine of subtilisin to —CH_2SH (Neet and Koshland, 1966; Polgar and Bender, 1966) and the conversion of the —CH_2COOH of the active-site aspartic acid of lysozyme to —CH_2CH_2OH (Eshdat *et al.*, 1974). It would be wrong to exclude completely such ingenious manipulations of protein structure merely on grounds of definition. We have seen, particularly in Chapter 5, that in addition to the interest of results obtained by using such analogues, the methods developed for such experiments have a great deal to offer to those undertaking the type of work discussed in this book. However, it is the use of the degradation products of proteins that we must consider most intensively.

We saw in Chapter 1 that the fragments could be combined either non-covalently or covalently. If covalent, the bonding can be by oxidative formation of disulphide bridges or by peptide bonds. Each of these three approaches has a specialized methodology of its own. While this Chapter concentrates on semisynthesis through peptide bonds (since I believe that it is the most generally applicable form) the importance of the many studies that have used non-covalent and oxidative methods demands that they too receive some mention.

NON-COVALENT SEMISYNTHESIS

As mentioned in Chapter 1 several of the proteins that are most often studied by molecular biologists show the phenomenon that, if they are cleaved into long

* Some workers prefer 'partially synthetic' and 'partial synthesis' as alternative terms.

fragments, certain mixtures of fragments will form non-covalently stabilized complexes which possess some of the properties of the intact parent protein. In particular pancreatic ribonuclease, *S. aureus* nuclease, cytochrome *c*, and somatotropin have this property and their non-covalent complexes have been exploited for the production of analogues. There is a list on p. 22 of the systems of this type; some have not yet been exploited for semisynthesis.

Pancreatic ribonuclease

The discovery (Richards, 1958) that pancreatic ribonuclease can be cleaved by subtilisin, and that the fragments form an active complex, has been vigorously exploited. The smaller of the fragments produced (the S-peptide) has been replaced by totally synthetic analogues (e.g. Finn and Hofmann, 1965; Hofmann *et al.*, 1966; Scoffone *et al.*, 1967; Moroder *et al.*, 1969; Chaiken *et al.*, 1973; Chaiken, 1974).

A large number of semisynthetic complexes have been prepared and have been studied in a wide variety of kinetic experiments. Other techniques, including X-ray crystallography and nuclear magnetic resonance, have also been used on them. Some analogues of the S-peptide bind to the larger fragment (the S-protein) but do not activate it: His^{12} has been shown by this means to be catalytically active and Phe^{8} and Met^{13} are important for binding between the S-peptide and S-protein. This work has been reviewed several times, most recently by Chaiken (1978). We have already seen (p. 123) that it has recently been found that the peptide bond between the S-protein and S-peptide can be enzymatically resynthesized.

Gutte and Merrifield (1971) obtained material from a study directed to the solid-phase synthesis of the S-protein which enabled them to demonstrate that the 5 residues 21 to 25 are not required for the binding of the S-peptide, or for enzymic activity.

Pandin *et al.* (1976) have studied the X-ray diffraction of crystals of a complex between a synthetic sample of the S-peptide (of the natural sequence) and a natural sample of the S-protein. They were unable to detect any differences to the patterns produced by crystals of the complex of entirely natural origin: they conclude that techniques now exist for preparing large quantities of highly purified semisynthetic analogues, susceptible to X-ray analysis.

Lin *et al.* (1970) produced a complex between enzymically degraded S-protein (6 residues were removed from the carboxyl terminus), the S-peptide, and a synthetic peptide corresponding to the last 14 residues of the native enzyme. The complex had about 30% of the activity of ribonuclease.

Staph. aureus *nuclease*

S. aureus (*Foggi strain*) produces an extracellular nuclease which has been the subject of intensive chemical and physical study. Under appropriate conditions, a limited tryptic digest produces only three fragments. The two larger fragments (one, P_2, corresponds to residues 6–48 and the other, P_3, (49 or 50)–149) form a non-covalent complex, nuclease-T, with about 8% of the activity of the native enzyme (Taniuchi *et al.*, 1967). The complex has been shown to be similar in structure to the native enzyme by a variety of techniques, including X-ray crystallography (Taniuchi *et al.*, 1972). Peptide P_2 has been synthesized by the solid-phase technique and, when highly purified, produces a semisynthetic nuclease-T of good quality. By contrast the crude synthetic product, as released from the solid-phase resin, gives a complex of low activity—approximately 0.4% of the activity of the native enzyme. Nonetheless, many semisynthetic analogues of nuclease-T have been prepared via solid-phase synthesis of P_2 and have given a great deal of information about the relationship between structure and activity. Of the analogues that have been prepared, those formed from the crude synthetic products were of low activity, but could be used qualitatively for identification of the residues that are essential for activity or the binding between P_2 and P_3. Such studies have in particular implicated Asp^{19}, Asp^{21}, Asp^{40}, Glu^{43}, and Arg^{35} in activity and have served to show that His^{46} is not catalytically important. These deductions are all consistent with the results of X-ray crystallography.

Di Bello (1975), in an effort to improve the quality of the analogues, carried out a pilot study for the covalent semisynthesis of P_2. His approach was to use the solid-phase technique to produce residues 36–47 of the sequence and then to complete the peptide by coupling on the appropriate natural fragment. This was a preliminary study, and ways in which the details of the approach used would be likely to be modified are discussed on p. 182. Even so the semisynthetic P_2 that he prepared was found to give a nuclease-T that was about three times as active as that made from a comparable, but totally synthetic, sample of P_2.

Another kind of active complex has been found (Taniuchi and Anfinsen, 1971). It consists of a tryptic fragment (residues 1–126) and one produced by CNBr (residues 99–149). Parikh *et al.* (1971) have used this system to produce analogues that have their substitutions in the carboxyl-terminal region.

Cytochrome c

Horse-heart cytochrome *c* has 104 residues, of which 2 (residues 65 and 80) are methionine and 2 (residues 38 and 91) are arginine (Margoliash *et al.*, 1961). Corradin and Harbury (1971) found that CNBr fragments 1–65 and 66–104 could be recombined after separation to form a functioning complex which was understandably thought to have been non-covalent. We now know (Corradin and Harbury, 1974a) that the broken peptide bond 65–66 re-forms spontaneously by ammonolysis of homoserine-lactone-65, and the practical implications have been extensively discussed in previous Chapters. The proper place to consider this useful discovery (and the analogous phenomena discovered in pancreatic trypsin inhibitor by Dyckes *et al.* (1974)) is in the section on fragment-condensation semisyntheses below. However, we should note that the complex formed between *apo*-cytochrome *c* (i.e. the protein minus the haem) and the haem peptide, 1–65 (Fisher *et al.*, 1973), must be a true non-covalent complex.

The limited tryptic digest produces a less ambiguously non-covalent complex. Of the two arginines, that at position 91 lies between two acidic residues, and probably as a result trypsin is reluctant to promote cleavage at this site. Furthermore, tryptic cleavage at lysine is prevented by acetimidylation (which preserves both the change on the lysyl side-chains and the biological activity of the protein; Harris and Offord, 1977). Therefore the only site of tryptic cleavage is the bond Arg^{38}-$Lys(N^{\epsilon}\text{-Acim})^{39}$ (Harris and Offord, 1977; Harris, 1978). Harris and Offord have found that the two peptides 1–38 and 39–104 can be recombined after separation to produce a functioning complex which resembles the intact protein quite closely in its visible, u.v. and n.m.r. spectra. They have also exploited this phenomenon to produce some semisynthetic analogues, in which Lys^{39} has been removed and replaced by stepwise methods by ornithine and by *p*-fluoro-L-phenylalanine. Harris (1978) has also described the removal of Arg^{38} from peptide 1–38 with carboxypeptidase B. It would be very desirable to try to find some means of re-forming the bond 38–39, since semisyntheses of analogues of the intact protein could then be carried out with very minimal schemes of protection.

Somatotropin

Li *et al.* (1978) have combined a natural amino-terminal fragment (residues 1–134) of human somatotropin with a synthetic analogue of residues 140–191 and in a separate experiment with an analogue of residues 145–191. The non-covalent complexes that were produced have full growth-promoting activity and nearly full immunoreactivity.

Myoglobin

Hagenmaier (1978) has reported the discovery of a complex between overlapping fragments of myoglobin. The fragments correspond to residues 1–55 of the deer protein (obtained by CNBr cleavage) and residues 32–139 of the horse protein (obtained after tryptic digestion limited to arginine residues by methylmaleylation). It seems that the remaining part of the sequence (residues 140–153) is not necessary for the combination. He describes some preliminary experiments directed toward the exploitation of this discovery.

Fragment F_v

Gavish *et al.* (1978) have produced a semisynthetic immunoprotein by non-covalent association between polypeptide fragments. Protein 315, an IgA molecule produced by a certain type of mouse myeloma, has a high affinity for the dinitrophenyl group. A proteolytic fragment, F_v, contains the full binding activity. The fragment consists of two chains, each about 110–120 residues long, held together by non-covalent forces. One of the two chains was prepared by solid-phase synthesis and the product extensively purified. The natural and synthetic fragments were allowed to associate and produced a semisynthetic molecule. The final yield of active material in a partly purified sample of product was 2%. This result constitutes a promising start on a project of great interest, but the usual problems of the conventional synthesis of long polypeptide chains are already apparent.

COVALENT SEMISYNTHESIS

The component parts of proteins are commonly joined by two types of covalent bond: the disulphide bridge and the peptide bond. Both types have been used in semisynthesis to join the fragments. It is possible to prepare analogues of more proteins by these methods than by using the non-covalent approach, and, once prepared, such analogues may well be more comparable in their behaviour to the natural form than are the non-covalent ones.

Covalent semisynthesis via disulphide bridges

Disulphide bridges are found almost exclusively in proteins that function outside cells. In most instances the bridges link different parts of a single polypeptide chain and are therefore of little use for linking synthetic to natural fragments when constructing an analogue. In such circumstances the most that they can do is hold the protein together if a cut is made at a site within the sequence as a part of a semisynthesis via peptide bonds (e.g. the work of Dyckes *et al.* (1974), discussed on p. 191).

However, some proteins, notably insulin, consist of more than one chain joined through disulphide bridges. (As usual it derives its two chains from a single, disulphide-bridged sequence after proteolytic cleavage.) Provided that the correct pattern of bridges can be formed, a natural and a synthetic chain in the reduced, cysteinyl form can be joined oxidatively to form a semisynthetic analogue (e.g. Scheme (8.1)).

A natural chain has been joined to a synthetic one of natural sequence on several occasions and there are numerous examples of a synthetic or semisynthetic analogue of one chain being combined with a natural sample of the

SH SH SH SH
+
SH SH
[O] →

S—S S—S
S S + S S
S S S S
correctly bridged product
incorrectly bridged products (13 other isomers)

(8.1)

other (see below). The A-chain of insulin from one species has been joined to the B-chain of insulin from another (e.g. Dixon, 1964).

The correct pattern of bridges is only one of 15 possibilities, even if we exclude patterns that link more than one chain of the same type. In order to maximize the yield of the correct product we must rely on the non-covalent forces that stabilize the tertiary structure of the native protein to bring the proper pairs of cysteine residues into proximity. An unnatural analogue may lack the proper balance of forces, with a consequent fall in the yield of correctly paired product, but even when the native sequences are used a mixture of products results. Some of the residues that had contributed their interactions to the formation of the correct pairs in the single-chain, precursor stage are lost by the proteolytic cleavages that gave rise to the two-chain, active form of the hormone. When a part of the original single chain has been lost there is no need *a priori* to assume that the correct pattern of pairs is more likely to form than any other possible pattern. Indeed, the conditions for the successful joining of the chains of insulin proved completely elusive for some years before painstaking attention to detail led to the elaboration of a procedure that would, using reduced chains from the natural protein give yields of 25–50% (Insulin Research Group, Shanghai, 1966) or, more usually, 7–15% (Zahn *et al.*, 1972).

Synthetic chains, even when of the natural sequence and made with great care, did not give nearly such good results. (Semisynthetic ones are better, but still not so good as natural ones.) The 1966 paper cited above speaks of yields of correctly bridged product falling to 5–10% when only one of the chains was natural and to 1.2–2.5% when both were synthetic. Although these last figures have been somewhat improved in more recent years, the recombination of chains is still a markedly inefficient process. Even if the correct bridges are formed, the tertiary structure may not be totally authentic. Brandenburg *et al.* (1972) found that the circular-dichroism spectrum of an acetyl derivative that they formed by the reoxidation of separated chains differed from the spectrum of the same derivative made by a process in which the disulphide bridges were intact throughout.

The efficiency of recombination in disulphide-bridged semisyntheses carried out in recent years has been reviewed by Geiger (1976). In every instance the recoveries of material are poor, and the authors concerned often explicitly attribute such findings to the change, mentioned above, which the introduction of an alteration into the natural sequence inevitably brings about in the balance of non-covalent forces.

In an attempt to improve this unsatisfactory situation Lindsay (1972), Brandenburg *et al.* (1973), Geiger and Obermeier (1973), and Busse and Carpenter (1974, 1976) have reported the replacement of the portion of the chain that is lost on proteolytic activation of the precursor by an artificial link between the α-NH_2 of the A-chain and the ε-NH_2 of the lysine residue adjacent to the carboxyl-terminal residue of the B-chain.

Such a link appears to supply much of that part of the influence toward proper pairing which is lost on the departure of the activation peptide (which joins the

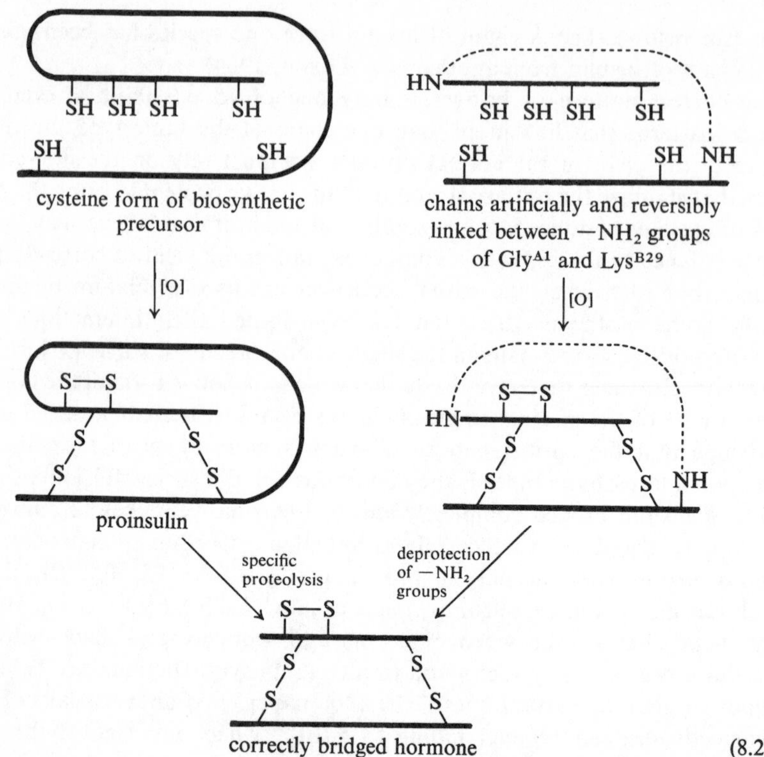

(8.2)

α-NH_2 of the A-chain to the α-COOH of the carboxyl-terminal residue of the B-chain). The percentage of correctly bridged product when natural chains are used rises to 60–75%. This type of approach, which has recently been reviewed by Carpenter *et al.* (1978), seems very likely to come into extensive use as a means for the preparation of semisynthetic and synthetic analogues.

Altogether, it seems best to avoid covalent semisynthesis via disulphide bridges at present, if at all possible. If it is forced on one then, inefficient though it can be, it is at least a source of modest quantities of useful materials that are not easily obtainable by any other means.

COVALENT SEMISYNTHESIS VIA THE FORMATION OF PEPTIDE BONDS

We can now consider the semisyntheses which represent the main subject matter of this book—those that produce analogues *via* the formation of normal peptide bonds between α-COOH and α-NH_2 groups. As we have observed so often before, both stepwise and fragment-condensation semisyntheses are possible.

Stepwise semisyntheses

The fore-runner of work of this type was the uncontrolled addition of chains of amino-acid residues to all the α- and ε-NH_2 groups of proteins. Stahman and

Becker (1952) allowed the —NH_2 groups to initiate the polymerization of *N*-carboxyaminoacyl anhydrides, which led to the formation of a mixture of $N^{\alpha,\epsilon}$-polypeptidyl derivatives of the protein. The preparation of polytyrosyl trypsin (Glazer *et al.*, 1962) is typical of a large number of such syntheses. Levy and Carpenter (1967) operated on insulin in a more controlled manner by the use of Boc-aminoacyl-*p*-nitrophenyl esters. Instead of a polypeptide chain of rather indeterminate length, just one residue was added to each of the α- and ε-NH_2 groups. Their derivatives included trialanyl, triasparaginyl, trilysyl, trimethionyl, and triglutamyl insulins.

The methods of Levy and Carpenter do not greatly differ in their details from those used subsequently for the addition of a single residue to a chosen α-NH_2 group. Such operations, in which there is complete control over the nature of the product, are now the dominant kind of stepwise semisynthesis. They have been carried out on insulin and the separate chains of insulin, clostridial ferredoxins, the functioning complex of the tryptic fragments of cytochrome *c*, phospholipase A_2, trypsin, and myoglobin. The common feature of this approach is, as previous Chapters led us to expect, that the protection of the —COOH groups of the amino component is unnecessary, and that all that is required is to distinguish between α- and ε-NH_2 groups. In the review of the stepwise work that follows, it will become apparent that there have been four major types of response to the need to distinguish α- from ε-NH_2 groups. In a few cases (the work on ferredoxin) the sequence of the protein is such that there is no need to make the distinction at all, and the problem can be ignored. If, as is true in most cases, the distinction must be made, then it can be done either by a chemical reaction specific to one type of group only (e.g. metal-catalysed N^{α}-transamination of trypsin), or by some less specific reaction. If the reaction is non-specific, we can rely on different reactivities of the groups, as has been done for some experiments on insulin, myoglobin, and glucagon, although it is then normal to have to find a separation method (e.g. ion exchange) to remove over- and under-reacted derivatives. Alternatively, the non-specific reaction can be taken to completion on a zymogen, which is subsequently activated to unblock the α-NH_2 group. This approach has been employed with trypsin and phospholipase A_2. Naturally, it calls for a form of —NH_2 protection that does not disrupt the conformation (and thus specificity of cleavage in the activation) of the zymogen.

Of the order of 100 analogues have been prepared in a sufficient state of purity to permit them to be characterized. We shall now examine the results in detail, protein by protein.

Insulin: methodology

Several workers have carried out amino-terminal modifications on insulin by stepwise semisynthesis. Usually, residues are first removed by the Edman procedure and then replaced by active-ester coupling, although the degradation step has occasionally been omitted. The operations have been carried out on both the A- and B-chains at once (Brandenburg and Ooms, 1968) or on separated chains (Weinert *et al.*, 1969; Shimonishi, 1970; Weinert *et al.*, 1971; Brandenburg *et al.*, 1975; Krail *et al.*, 1975a; Cosmatos *et al.*, 1978), but it is probably best to leave the disulphide bridges intact and exploit the selective reactivity of one or other of insulin's α-NH_2 groups in order to produce modifications in one chain only (Borrás and Offord, 1970a, b; Krail *et al.*, 1971; Saunders and Offord, 1972; Geiger *et al.*, 1975b, 1976; Halban and Offord, 1975; Krail *et al.*, 1975b; Teetz *et al.*, 1976; Saunders and Offord, 1977a, b; Geiger *et al.*, 1978; Saunders, 1978). Schemes (8.3) and (8.4) show these two possible approaches in outline. The uses to which the analogues have been put are discussed later in this section.

(8.3)

The details of the methods used for these semisyntheses have changed with time in a manner which reflects the general evolution of peptide synthesis during the period in question. At the beginning, a wide range of protecting groups, solvents, bases, and even types of activation were employed. More recently, matters have settled down to the use of one or two active esters of Boc-amino acids. The *N*-hydroxysuccinimido esters have been the most popular, though the 2,4,5-trichlorophenyl esters are also used quite often. Aqueous or semi-aqueous solutions are less often used than they were, which reflects a greater confidence in the ability of organic solvents to dissolve insulin and its chains.* The peptide chemist's traditional distrust of dimethylsulphoxide as a solvent has been overcome and, as in many conventional systems, *N*-alkylmorpholines are being used as bases where once compounds such as triethylamine would have been the natural choice.

The most crucial developments in technique are probably those that have concerned the exploitation of the specific reactivity of the —NH_2 groups of

* We now take for granted our ability to dissolve proteins in organic solvents, and in strong anhydrous acids and bases, without their being damaged. Yet this realization, which has been brought about by semisynthetic work probably as much as by any other factor, is not a trivial one and has considerable technological implications.

insulin. Ways have been found that make it possible to operate on one chain at a time without having first to break the bridges and separate them. An early method was to exploit the fact that, of the three —NH_2 groups, that of residue **B**1 has the greater reactivity for phenyl isothiocyanate (Borrás and Offord, 1970a, b; and see Example (2.7)). In contrast, all the semisyntheses that have been reported more recently have, if the disulphide bridges were left alone, involved a

H_2N—Gly— (S—S) ... H_2N— ... NH_2 —selective protection→

Boc—NH—Gly— (S—S) ... H_2N— ... NH_2 —further acid-stable protection→

Boc—NH—Gly— (S—S) ... R—NH— ... NH—R —acid treatment→

H_2N—Gly— (S—S) ... R—NH— ... NH—R —removal and replacement of amino-terminal residue→

H_2N—X— (S—S) ... R—NH— ... NH—R —final deprotection→

H_2N—X— (S—S) ... H_2N— ... NH_2

R = an acid-stable protecting group (8.4)

stage in which advantage was taken of the greater reactivity of the —NH_2 groups of glycine A1 and lysine B29 towards certain protecting reagents. So marked and so convenient is this preference that it has been exploited even when modifications are wanted in the A-chain (Scheme (8.5)).

The selective reactivity of the —NH_2 groups of residues A1 and B29 was first observed with t-butyloxycarbonyl azide (Geiger *et al.*, 1971; see Example (2.8)). Under the conditions of Geiger *et al.* the N^{α}-Gly^{A1},N^{ϵ}-Lys^{B29} di-derivative is the principal one, although it must be separated from the small quantity of mono- and tri-Boc- derivatives that contaminate it. The resulting purified

$$N^{\alpha}\text{—Gly}^{A1}, N^{\epsilon}\text{—Lys}^{B29}\text{—di-Boc-insulin}$$

$$\downarrow$$

$$N^{\alpha}\text{—Gly}^{A1}, N^{\epsilon}\text{—Lys}^{B29}\text{—di-Boc}, N^{\alpha}\text{—Phe}^{B1}\text{-Msc-insulin}$$

$$\downarrow$$

$$N^{\alpha}\text{—Phe}^{B1}\text{-Msc-insulin}$$

$$\downarrow \text{selective reaction}$$

$$N^{\alpha}\text{—Gly}^{A1}\text{—phenylthiocarbamoyl } N^{\alpha}\text{—Phe}^{B1}\text{-Msc-insulin}$$

$$\downarrow$$

$$N^{\alpha}\text{—Gly}^{A1}\text{—phenylthiocarbamoyl}, N^{\alpha}\text{—Phe}^{B1}, N^{\epsilon}\text{—Lys}^{B29}\text{—di-Msc-insulin}$$

$$\downarrow H^{+}$$

$$N^{\alpha}\text{—Phe}^{B1}, N^{\epsilon}\text{—Lys}^{B29}\text{—di-Msc-des-Gly}^{A1}\text{-insulin}$$

$$\downarrow \text{coupling with } N^{\alpha}\text{ Boc—X—active ester}$$

$$N^{\alpha}\text{—Phe}^{B1}, N^{\epsilon}\text{—Lys}^{B29}\text{—di-Msc}, N^{\alpha}\text{—X}^{A1}\text{—Boc[X}^{A1}\text{]insulin}$$

$$\downarrow \text{base}$$

$$N^{\alpha}\text{—X}^{A1}\text{—Boc[X}^{A1}\text{]insulin}$$

$$\downarrow H^{+}$$

$$\text{[X}^{A1}\text{]insulin} \qquad (8.5)$$

di-derivative was found to be substantially free of the other two possible di-Boc-products, which is fortunate since they would have been harder to remove. Phenylalanine B1 can then be removed by the Edman process, as described in Example (3.1). Since the Boc- group is labile to the acid step of the Edman procedure (Reaction (3.2)), the —NH_2 protection is lost. The product is des-Phe^{B1}-insulin, first prepared by another route by Brandenburg (1969). Fortunately, the Boc- group can be reapplied to residues A1 and B29 of the des-Phe^{B1}-insulin with similar degree of selectivity to that seen with the intact molecule. After purification as before, the di-Boc- derivative was ready to act as an amino component in the preparation of analogues altered at B1 or for further selective degradation to des-phenylalanine-valine (B1–B2)-insulin. Geiger and Langner (1973) prepared by this means the two truncated insulins already mentioned and, by another cycle of protection, purification, and degradation, the des-phenylalanine-valine-asparagine-(B1–B3)-insulin. The amino-terminal residue of the B-chain of this last derivative, glutamine B4, was found to be cyclized to pyroglutamic acid. Geiger and Langner do not report whether or not the parent compound of the des-(B1–B3)-insulin (the N^{α}-Gly^{A1},N^{ϵ}-Lys^{B29}-di-Boc-des-(B1–B2) derivative) was free of its other two possible isomeric di-Boc- derivatives.

It is clearly wasteful of effort to be continually deprotecting the same two groups and each time to have to reprotect and purify the product. In an attempt to overcome this drawback, Geiger and Langner (1975) applied the phthaloyl group, which is stable to acid (p. 36). Conditions could be found in which it, too, showed a preference for A1 and B29. Geiger and Langner carried out five cycles of the Edman reaction on the purified di-derivative this time without having to reprotect after each cycle. Unfortunately the des-phenylalanine-valine-asparagine-glutamine-histidine-(B1–B5)-insulin did not withstand the N_2H_4 treatment needed to remove the group. Even the presence of phenol failed to prevent the formation of side-products, although it had done so during deprotection of the less extensively truncated phthaloyl insulins. More recently Geiger *et al.* (1975a) have reported the use of the methylsulphonyl-ethyloxycarbonyl group, which is stable to acid. It is labile to relatively strongly basic conditions but these need only be applied for very short periods (e.g. pH 13 for 30 s).

Saunders and Offord ((1977a, b); and see Example (2.4)) have used the benzyloxycarbonylmethionyl- and ethyloxycarbonylmethionyl- groups for a similar purpose. These groups are not removed by the acid step of the Edman reaction. They are, therefore, also of particular value for use when repeated, stepwise shortening of the chain is intended.* In addition to the analogues made by removal and replacement of residues, both the $CNBr^-$ and the base-labile types of group have been used for the production of the complete series of insulins truncated in the B-chain up to and including the des-(B1–B5)-insulin (Saunders and Offord, 1977a; Geiger *et al.*, 1978).

Apart from the studies already mentioned, a number of authors have reported work on the chemical modification of insulin that has particular relevance to stepwise semisynthesis. Levy (1973) has reported methods for the preparation of insulin with Boc- protection on any one of the three —NH_2 groups, as well as the N^{α}-Gly^{A1},N^{ϵ}-Lys^{B29} di-derivative of Geiger *et al.* (1971) and the tri-derivative of Levy and Carpenter (1967). Paselk and Levy (1974) have prepared all the possible mono- and di-trifluoroacetyl insulins, as well as the tri-derivative. They used phenyl trifluoroacetate, as Borrás and Offord (1970a, b) had, and ethyl trifluoroacetate (cf. p. 35). Finally, we should not forget the careful preparation of a number of specifically *N*-acetylated insulins by Brandenburg *et al.* (1972), which seem a great deal more semisynthetic if we reflect that the acetyl group could be described as a des-amino-glycyl residue.

Insulin: the use of the products

This is not a treatise on the relationship between structure and function in insulin and so I propose only to say sufficient about the experimental use of the results obtained as is necessary to put the syntheses in perspective.

* Saunders and Offord report that a small proportion of the Cbz- group is lost during the acid cleavage. They recommend the ethyloxycarbonyl group when repeated Edman degradations are planned.

The modifications carried out so far have been rather conservative, although the successful removal of the several residues of both the A- and B-chains might lead us to expect to see alterations made fairly soon deep in the molecule. The most striking conclusions concern the critical role of residue A1 in biological activity (Brandenburg *et al.*, 1972, 1975; Krail *et al.*, 1975a; and Geiger *et al.*, 1975); the discoveries that have been made would not have been possible without the semisynthetic technique. Geiger *et al.* (1978) describe other semisynthetic operations, which have thrown light on the roles of particular residues in both the A- and B-chains.

In contrast what was found at position A1, the relative unimportance of position B1, established through semisynthesis, makes it an ideal place to make substitutions that fit the resulting analogue for study by modern physical techniques (Krail *et al.*, 1971; Borrás, 1972; Saunders and Offord, 1972, 1977a; and see Geiger, 1976; Geiger *et al.*, 1978). The derivative reported by Saunders and Offord (1972)—[Gly^{B1}-^{13}C]insulin—gave clear n.m.r. signals (Fig. 8.1) which enabled the existence of one of the particular structural features of the hormone in the crystal to be inferred in the structure in free solution. Some others of this type of derivative at B1 may prove useful now that it has been shown that residue B1 is strongly involved in the transition from the 2-Zn to the 4-Zn form of insulin (Schlichtkrull, 1958). The 4-Zn form is the basis of several long-acting preparations of the hormone in widespread clinical use. Halban and Offord (1975) and Halban *et al.* (1976) discuss reasons why the semisynthetic [Phe^{B1}-^{3}H]insulin analogue should be more authentic in its behaviour than the more conventional radioactive insulins made by side-chain iodination. Tritium labelling is, also, intrinsically more attractive for some experiments *in vivo* and *in vitro*. The two papers cited describe the behaviour of the [^{3}H]insulin in a number of systems *in vitro* and *in vivo* together with methods for the analogue's long-term storage (a problem not previously encountered with the shorter-lived iodinated forms) and purification. This analogue has now begun to be employed in studies on the metabolism of insulin in both animals and man (e.g. Halban *et al.*, 1978).

Most of the authors cited in the previous few pages discuss the effect of the bulk and polarity of their substituent side-chain on the biological activity, immunological activity, and crystallizability of the hormone. In addition, the tertiary structure of several of the analogues has been studied by circular dichroism (e.g. Brandenburg *et al.*, 1972, 1973, 1975; Krail *et al.*, 1971, 1975a, b; Berndt *et al.*, 1975).

We shall see on p. 174 that further insights have been gained into structure and function in insulin by the use of semisynthetic analogues made by fragment condensation.

Clostridial ferredoxin

The ferredoxin of *Clostridium acidi-urici* is particularly suitable for stepwise semisynthetic work. It has only 55 residues and, since it lacks lysine (the sequence was determined by Rall *et al.* (1969)) has only one —NH_2 group, that of the terminal alanine. These facts have been energetically exploited. Hong and Rabinowitz (1970), apart from more traditional types of side-chain modification, coupled amino acids to Ala^1 to produce glycyl-, phenylalanyl-, lysyl-, glutamyl-, and methionyl-apoferredoxin. (It might be clearer to call such derivatives $[Gly^0]$apoferredoxin, and so on, but this would be contrary to the established rules of nomenclature.) The apoferredoxin was prepared by removal of the iron–sulphur clusters which are concerned in the biological activity, and the resulting free cysteinyl —SH groups protected by oxidation to disulphide bridges. Hong and Rabinowitz followed Levy and Carpenter (1967) in their use of the *p*-nitrophenyl esters of Boc- amino acids, but appear to have been obliged to use a semi-aqueous solvent. The products were deprotected and reductively combined with iron and sulphide to give the ferredoxin analogues. The side-chain of the added glutamic acid was deprotected by saponification, but the authors do not report to what extent, if any, Reaction (1.20) or an analogue of Reaction (7.1) might have taken place. The authors discussed the results of studies on the biological activity of the analogues and on the degree of stability of their binding to the iron–sulphur clusters.

Lode (1973) used the full semisynthetic procedure of Edman degradation followed by coupling to prepare the $[Gly^1]$, $[Leu^1]$, $[Phe^1]$, $[Met^1]$, $[Pro^1]$, $[Lys^1]$, and $[Glu^1]$apoferredoxins. He also removed the second residue (tyrosine) and then prepared des-Ala^1-$[Gly^2]$apoferredoxin, des-Ala^1-$[Phe^2]$apoferredoxin, and des-Ala^1-$[Trp^2]$apoferredoxin. (The original paper, which is a conference abstract, omits des-Ala^1- from the names of the analogues at position 2, but it seems clear from the context that Ala^1 had not been replaced.) Lode described the critical effect of certain of these deletions and substitutions on the biological activity and stability of the reconstituted ferredoxin analogues.

Tyrosine-2 is one of the only two aromatic residues in the sequence (the other is tyrosine-30), and the aromaticity of both these positions is normally conserved among the bacterial ferredoxins. The side-chains are close to the iron–sulphur clusters. Lode *et al.* (1974a) prepared des-Ala^1-Tyr^2-apoferredoxin in the usual way and then placed leucine in position 2 and alanine in position 1 to give $[Leu^2]$apoferredoxin. The coupling was carried out by the methods of Hong and Rabinowitz (1970). The same approach was used to make the $[Gly^2]$ analogue. The results of their examination of the properties of the reconstituted ferredoxin analogues, taken together with earlier work, suggested that only bulk, and not specifically aromaticity, was necessary. The bulk appears to be needed simply to maintain the stability of the binding to the iron–sulphur clusters.

Packer *et al.* (1975) have extended these studies to the semisynthetic $[Phe^2]$ analogue. They compared the nuclear magnetic resonance spectrum and other properties of this analogue with those of the native ferredoxin of *Cl. pasteurianum*. This latter protein has a tyrosine at position 2 and a phenylalanine-30 (Tanaka *et al.*, 1966), and thus has the converse distribution of aromatics to that of the $[Phe^2]$ analogue.

Lode *et al.* (1976a, b) have reported further studies on some of the analogues already mentioned, together with the [His[2]], [*o*-fluoro-L-Phe[2]], [*m*-fluoro-L-Phe[2]], [*p*-NH_2-Tyr[30]], and [Leu[2], *p*-NH_2-Tyr[30]] analogues. All of these derivatives were prepared via the des-Ala[1]-Tyr[2]-apoferredoxin. This intermediate contains only the one tyrosine, at position 30, and this fact was exploited to introduce the $-NH_2$ group into the side-chain of this residue by nitration of the truncated intermediate and reduction of the nitro group subsequent to the replacement of residues 1 and 2. Some of the analogues were made by what is strictly fragment-condensation semisynthesis in that both residues were replaced at once by condensation with the *N*-hydroxysuccinimido ester of the appropriate Boc-dipeptide. The particular procedure described does not completely exclude the possibility of some racemization during coupling (cf. p. 7), but the authors do not describe any tests on their analogues that would enable one to assess to what extent, if at all, it had occurred.

Lode *et al.* (1974b) have carried out semisynthetic work on the ferredoxin of *Clostridium m-e*. Their approach was such as to make it more sensible to discuss this paper in the section on fragment-condensation semisynthesis (p. 180).

Trypsin

The major component of bovine cationic trypsinogen has the amino-terminal sequence Val-Asp-Asp-Asp-Asp-Lys-Ile-,* and 15 of its 229 residues are lysines (Walsh and Neurath, 1964; Mikes *et al.*, 1966).

The zymogen is activated by proteolytic cleavage of the Lys-Ile bond in the amino-terminal sequence shown above. The positive charge of the newly liberated α-NH_2 group of the isoleucine residue is thought to cause structural changes in the region of the enzyme's active centre that are important in the acquisition of catalytic activity. In spite of the large number of ε-NH_2 groups Webster and Offord (1972) and Webster (1974) sought to use stepwise semisynthesis to contribute to the studies on this question. Webster and Offord used metal-catalysed transamination of the α-NH_2 group as a temporary means of protection (Reaction (3.3)) because it is absolutely specific for α- as opposed to ε-NH_2 groups. They protected the ε-NH_2 groups by acetimidylation and removed the transaminated residue by means of Reaction (3.4). Webster (1974) replaced the isoleucine. A puzzling loss of solubility in the early stages of the route hampered further work, but these problems do not occur with every protein (Dixon and Fields, 1972). In spite of the difficulties encountered (see also p. 52) transamination should not be forgotten.

Robinson *et al.* (1973) have carried out some preliminary experiments directed toward the semisynthesis of phenylalanyl-trypsin.

* As a result of this accumulation of activation peptide as a by-product of a semisynthesis, Webster and Offord (1977) noticed a minor component with two additional residues (Phe-Pro-) at the amino terminus. It seems likely that both forms are found in a single animal (Mahony and Offord, 1977).

Acetimidylated trypsin is fully active (Nurredin and Inganami, 1969, 1975) and this presents an opportunity for fragment-condensation work at the active-site residue, serine-183. The inhibitor phenylmethanesulphonylfluoride acts by acylating the —CH_2OH group of this residue, but if removed in acid solution, induces an N→O acyl shift of the —COOH group of residue 182 (Fahrney and and Gold, 1963; Gold and Fahrney, 1967).

$$-\overset{\overset{\displaystyle O}{\|}}{C}-NH-\overset{\overset{\displaystyle CH_2OH}{|}}{CH}-CO- \;+\; C_6H_5-CH_2-SO_2F \longrightarrow$$

$$-\overset{\overset{\displaystyle O}{\|}}{C}-\underset{H}{N}-\overset{\overset{\displaystyle H_2C-O-\overset{O}{\overset{\|}{S}}(=O)-CH_2-C_6H_5}{|}}{CH}-CO- \xrightarrow{H^+} \overset{O\text{———}CH_2}{C\overset{+}{=}\underset{H}{N}-CH-CO-} \xrightarrow{H_2O} \overset{O\text{———}CH_2}{\underset{}{CO}\quad \underset{H_3\overset{+}{N}}{}\;CH-CO-} \qquad (8.6)$$

The resulting ester bond is the only one in the protein and could be cleaved specifically. If a mono-protected hydrazine were to be used then by analogy with Reaction (6.2) the bond could be resealed after the stepwise removal and replacement of residue 183.

Phospholipase A_2

The precursor form of porcine phospholipase A_2 has 130 residues, of which the amino-terminal one is pyrrolidone carboxylic acid and 9 are lysine (de Haas *et al.*, 1970a). It has seven disulphide bridges (de Haas *et al.*, 1970b). The large number of lysine residues, and the lack of an amino terminus, made it seem at first sight a most unlikely candidate for a stepwise semisynthesis, but the situation becomes much more favourable when the precursor is activated.

The precursor is converted to the normal form of the enzyme by a proteolytic cleavage at arginine-7 (de Haas *et al.*, 1971). There results only a doubling of activity toward monomeric substrates (that is, dilute solutions of 3-*sn*-phosphoglycerides) but there is a very great increase in activity toward micellar aggregates of the substrate (de Haas *et al.*, 1971). The new α-NH_2 group (of alanine-8) is thought to be critically involved in the conformation changes that result in the improved ability to attach micelles, and Slotboom and de Haas (1975) and Slotboom *et al.* (1977, 1978) have prepared analogues with alterations in this region by stepwise semisynthesis. Like Webster and Offord (1972), they saw the advantages of the acetimidyl group. They acetimidylated the ε-NH_2 groups of the precursor and then activated it by limited tryptic hydrolysis between residues 7 and 8. End-group analysis showed that the other arginyl peptide bonds had not been affected to a detectable extent. The N^{ϵ}-acetimidyl enzyme had satisfactory biological activity and the authors did not, therefore, subsequently need to avail themselves of the option of removing the N^{ϵ}-protection (the deprotection gives a fully active product (Abita *et al.*, 1972)).

The N^{ϵ}-acetimidyl enzyme was modified in various ways. The des-Ala^8-N^{ϵ}-acetimidyl enzyme was prepared by the Edman process as were, in successive cycles, the des-Ala^8-Leu^9- and the des-Ala^8-Leu^9-Trp^{10}- derivatives. Amino-acid residues (and other groups) were attached to the α-NH_2 of the N^{ϵ}-acetimidyl enzyme and to those of its truncated derivatives. The hydroxysuccinimido esters of Boc-amino acids were used in a semi-aqueous solvent to prepare the chain-elongated Ala^7-, DL-[^{13}C]Ala^7-, and [^{13}C]Gly^7- analogues by reaction with the N^{ϵ}-acetimidyl enzyme. The [β-Ala^8], [Gly^8], [Leu^8], [Asn^8], [Phe^8], [L-[^{13}C]Ala^8], [D-[^{13}C]Ala^8], [^{13}C-Gly^8] analogues were prepared in a similar manner by reaction with the des-Ala^8- enzyme. These couplings were found to result in the incorporation of more than one residue of amino acid into the protein. However, since only one residue survived hydroxylamine at neutral pH (a treatment which did not seem to harm the protein) the extra degree of incorporation was put down to the formation of aminoacyl esters with side-chain —COOH groups (Slotboom *et al.*, 1978). The hydroxysuccinimido ester of Boc-Ala-Leu-Phe was coupled to the des-Ala^8-Leu^8-Trp^{10} derivative to give the [Phe^{10}] analogue. (The same possibility of a small degree of racemization might have arisen here as it did in the ferredoxin semisynthesis.)

A great many experiments have been carried out on these analogues. The results have contributed significantly to the identification and characterization of the part of the protein that, in distinction to the ordinary catalytic and specificity sites, recognizes the presence of an interface between water and lipid. Apart from testing the analogues against both substrate micelles and substrate in free solution, extensive use has been made of a number of physical techniques. These include ultraviolet difference spectroscopy; fluorescence spectroscopy; measurements of the influence of surface pressure on the lag-time of attack by the enzyme on substrate in monolayer form; and (for the ^{13}C-labelled analogues) nuclear magnetic resonance. Since the interface-recognition site was found to be in the region subjected to semisynthetic manipulation, these experiments have proved very informative. The site depends critically on a salt bridge of which one partner is the α-NH_2 group of residue, and so pH titrations, together with a study of the effect of Ca^{2+} and other ions, have proved particularly useful. The response of the ^{13}C nuclear magnetic resonance peaks to changes in pH have made it possible to measure pK values of the α-NH_2 group and variation in pK among analogues has been correlated with the alterations in structure between one analogue and another. There can be few proteins in which there have been made so many small, precise changes in structure, and which have yielded so much reliable information as a result.

The calcium ion (Ca^{2+}), which binds to the protein, can be replaced by Gd^{3+}. This latter ion is paramagnetic, and its effect on the ^{13}C resonances of [L-[^{13}C]Ala^8] and [D-[^{13}C]Ala^8] was used to measure the distance between the metal-binding site and the amino terminus.

An additional finding has opened the way to true fragment-condensation semisyntheses a little deeper into the chain (as well as to more stepwise semisyntheses). Slotboom and de Haas (1975) have discovered that if tryptic cleavage is continued after the activation, the peptide bond between Arg^{13} and Ser^{14} is cleaved very much more readily than the three other arginyl peptide bonds that remain, further along the sequence.

Myoglobin

The amino-terminal residue of sperm-whale myoglobin is glycine; 19 of the 153 residues are lysine (Edmundson, 1965). Garner *et al.* (1973) investigated the coupling of [^{13}C-Gly^2] to the α-NH_2 group of the unprotected protein in order to carry out nuclear magnetic resonance studies. Garner and Gurd (1975) improved the specificity of coupling by prior acetimidylation of the protein. They obtained the pK of the new α-NH_2 group of their product by following the change in chemical shift of the resonance of the ^{13}C atom with pH.

Methyl acetimidate shows some preference for ε-NH_2 groups over α-NH_2 groups and, although apparently not homogeneous, the product was predominantly [1-^{13}C]glycyl myoglobin. Because of the differences in pK between α-NH_2 groups, ε-NH_2 groups, and acetimidyl groups it is possible to separate side-products from the wanted intermediate, fully N^{ε}-acetimidylated myoglobin (Di Marchi *et al.*, 1978a). This purified intermediate has been used to make a more homogeneous chain-elongated product, and as the starting point of a controlled stepwise degradation for semisynthesis. Di Marchi *et al.* (1978a) made [^{13}C]glycyl myoglobin from the purified N^{ε}-acetimidyl derivative. The glycine was introduced as the N^{α}-trifluoroacetyl hydroxysuccinimido ester. They also used 3-sulphophenyl isothiocyanate to excise Gly^1—the sulpho group facilitated

$$H_2N{-}CH_2{-}CO{-}NH{-}\underset{}{\overset{R_1}{\overset{|}{C}H}}{-}CO{-} \longrightarrow H{-}CO{-}CO{-}NH{-}\overset{R_1}{\overset{|}{C}H}{-}CO{-}$$

$$H{-}CO{-}CO{-}NH{-}\overset{R_1}{\overset{|}{C}H}{-}CO{-} + \underset{\text{L-glutamic acid}}{H_2N{-}CH(CH_2CH_2COOH){-}COOH} \longrightarrow$$

$$H_2N{-}CH_2{-}CO{-}NH{-}\overset{R_1}{\overset{|}{C}H}{-}CO{-} + CO(CH_2CH_2COOH){-}COOH \qquad (8.7)$$

the chromatographic removal of uncleaved material—and then replaced it with [^{13}C]Gly. Both [1-^{13}C]- and [2-^{13}C]glycine were used in the two sets of experiments to give four derivatives in all. After checks that the products had retained or regained the authentic tertiary structure (by circular dichroism and, with reference to the position of the haem, by visible-light spectroscopy) they were subjected to n.m.r. p*K* values were obtained for the α-NH_2 group and compared with values obtained by other means.

In spite of the drawbacks noted on previous pages, it might also be worthwhile to try metal-catalysed transamination. In particular, glycine (which, as we have seen, is the amino-terminal residue) is unique among amino-acid residues in that it can be regenerated from the α-keto derivative formed on transamination (Dixon and Weitkamp, 1962; Reaction (8.7)). The temporary specific protection of the α-NH_2 group as the α-keto derivative would then be particularly convenient, because there would be no need to remove the residue by the rather uncertain Reaction (3.4) and replace it.*

We shall see later that a promising start has been made by several workers on fragment-condensation semisyntheses of myoglobin analogues.

Cytochrome c

Harris and Offord (1977) and Harris (1978) have, as already mentioned, exploited the existence of a functioning complex between two tryptic fragments of the protein (corresponding to residues 1–38 and 39–104) to prepare analogues. They have so far reported the preparation of the [*p*-fluoro-L-Phe^{39}] and [Orn^{39}] analogues of the complex.

Fragment-condensation semisynthesis

We have discussed the principles of fragment-condensation semisynthesis in previous Chapters. We saw that the major conceptual problem was to obtain a differential protection between the α-NH_2 and —COOH groups and their side-chain counterparts, and that this problem was solved in principle by Goldberger and Anfinsen (1962) for —NH_2 groups and by Offord (1969a, 1972) for —COOH groups. We must now consider how these proposals have worked out in practice, together with ingenious alternatives of less general application, such as those of Sealock and Laskowski (1969), in which proteases are used as the coupling agents, and of Obermeier and Geiger (1976) and Obermeier (1978) in which coupling is carried out without side-chain protection of the —COOH groups of the carboxyl component.

Insulin

There are only two sites for tryptic cleavage in insulin, at Arg^{22} and Lys^{29}. If we acylate the —NH_2 groups, we restrict tryptic cleavage to Arg^{22}. The

* Reaction (8.7) provides a means of inserting a ^{15}N atom into the α-NH_2 group of the protein. This isotope gives n.m.r. signals, but they are weak and in order to make use of the analogue it would probably require more sensitive instruments than are at present available.

octapeptide that is released, Gly-Phe-Phe-Tyr-Thr-Pro-Lys-Ala, corresponds to residues B23–B30 and is thus the carboxyl-terminal fragment of the B-chain. This region is of interest in that it contains a number of residues that tend to be conserved between the insulins of different species. Such an observation always suggests that the residues in question may have an important role in the protein, and in the present instance this impression is reinforced by the results of studies such as X-ray crystallography and the examination of analogues made by conventional synthesis, as well as by studies on side-chain derivatives (De Meyts *et al.*, 1978).

The scheme of —NH_2 protection, tryptic cleavage, —COOH protection, and selective deprotection by trypsin (Offord, 1969a) is therefore particularly attractive as applied to insulin (Fig. 8.1). There are only two fragments to couple, and analogues of the smaller of them, which contains sites of major interest, are relatively easy to synthesize. A number of laboratories have sought to exploit this situation, and it has become apparent that while quite unambiguously successful with some other proteins, the application of this scheme of reactions to insulin presents considerable difficulties.

Ruttenberg (1972) reported the use of the above reaction sequence in the semisynthesis of human insulin from the porcine des-octapeptide-(B23–B30)-insulin and a synthetic octapeptide corresponding to the human sequence. Internal evidence in the paper suggests that the claim was premature. To the extent that they were sufficiently explicitly described to be repeated, Ruttenberg's procedures have not led to his stated result in this Laboratory or elsewhere (Weitzel *et al.*, 1976; and see Discussion to Obermeier, 1978).

The Insulin Research Group, Shanghai (1973), used much the same overall route to prepare semisynthetic insulin. They prepared N^{α}-Gly^{A1}, N^{α}-Phe^{B1}-di-Boc-des-octapeptide-(B23–B30)-insulin penta-methyl ester and coupled it to phenylalanine methyl ester, and the methyl esters of the appropriate sequence to prepare, after deprotection, [Phe^{23}]-des-heptapeptide-(B24–B30)-insulin, des-hexapeptide-(B25–B30)-insulin, [Ala^{23}-Ala^{24}-Ala^{25}-]-des-pentapeptide-(B26–B30)-insulin, and insulin itself. They applied methyl protection to the natural fragment with methanol/HCl. (Ruttenberg (1972) had used diazomethane, although Offord (1969a) and Hayward and Offord (1971) had warned of the tendency of diazoalkanes to alkylate the unprotected side-chain of insulin.) They were obliged to remove the methyl group by saponification. Saponification poses the danger that Reaction (1.20) might occur, and also of amide loss, and there are general drawbacks to exposing proteins to the conditions used by the Insulin Research Group (0.1 M NaOH at 15 °C for 35 min). Therefore, in the absence of chemical characterization, the biological activity of their product is of particular interest. The crude resynthesized insulin had very low activity in the mouse-convulsion assay on small groups of animals, but after purification the specific activity rose somewhat. The semisynthetic des-hexapeptide-insulin also showed a certain amount of activity in this test, although in both cases the detailed, numerical values are a little hard to assess. The same paper illustrates the effect on blood-sugar levels in the rabbit (the

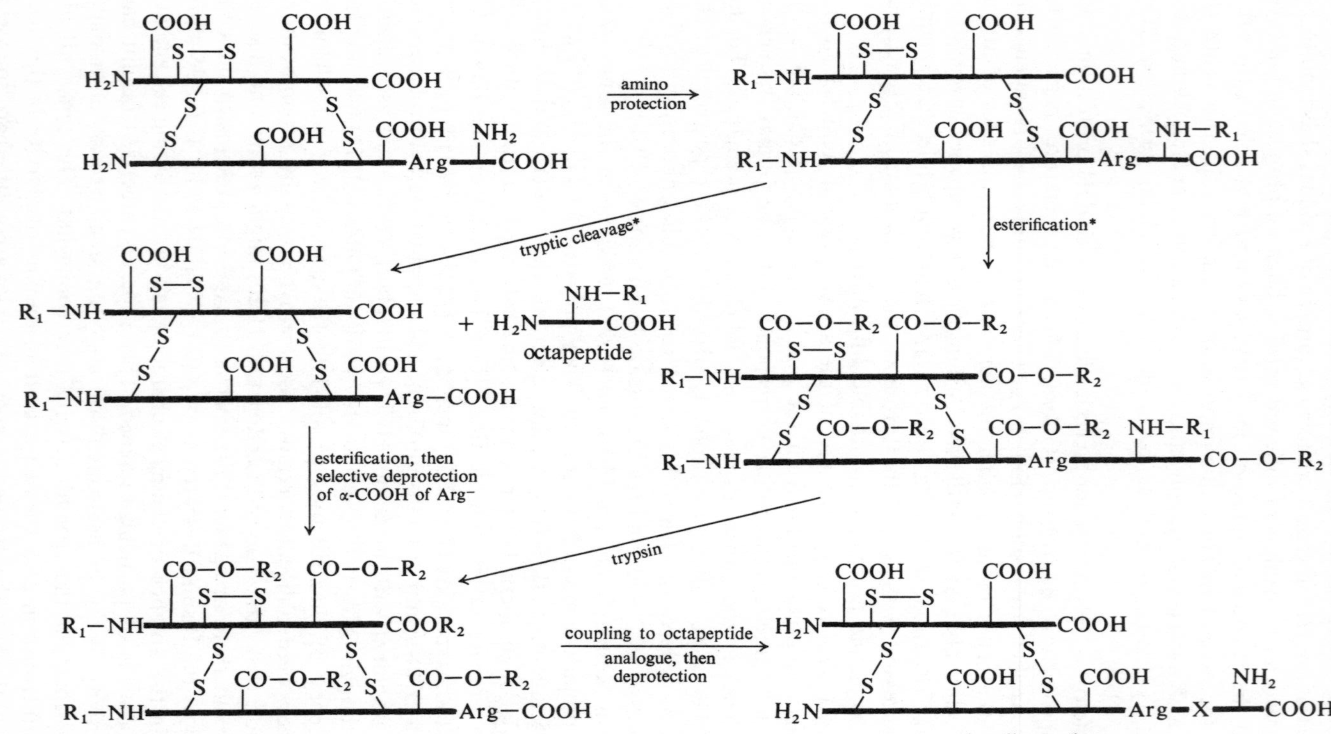

*These are alternative patterns.

Fig. 8.1. A route to B-chain analogues of insulin (see text).

number of animals is not stated) which, if we assume that the same quantities of material were used for each test, show that the des-hexapeptide derivative has more activity than the des-octapeptide material. It is not possible to assess the specific activity of these derivatives because the blood-sugar profiles for natural and resynthesized insulins did not rise from the post-injection minimum value before the ending of the experiment (or death of the animals ?) 4 h later.

The Insulin Research Groups of Pekin and Shanghai (1974) reported semisyntheses, by the same methods as in the 1973 paper, of des-hexapeptide-(B25–B30)-insulin, des-heptapeptide-(B24–B30)-insulin, and the [L-Ala^{23}] and [D-Ala^{23}] analogues of the des-heptapeptide-(B24–B30)-insulin. The des-hexapeptide-insulin and its [D-Ala^{23}] analogue possessed definite biological activity, although the mode of assay and the number of animals are not stated. The des-heptapeptide-insulin and the [L-Ala^{23}]-des-hexapeptide-insulin had little or no biological activity. The two active analogues were found to have apparent binding constants for the receptor of liver-cell membranes 125 and 70 times less intense, respectively, than that of native insulin. The two other analogues display very little specific binding at all.

Although these results may have told us something of the nature of the receptor-binding site on the insulin molecule, they are not such as to enable us to assess the reliability of the particular semisynthetic route employed.

Gattner *et al.* (1976) have also explored the general scheme and have studied the saponification step with great care (Gattner *et al.*, 1978). They showed that saponification of insulin hexamethyl ester gives a product that, as a consequence of Reaction (1.20), contains a considerable quantity of [isoasparagine A21]-insulin. This analogue has practically full biological activity (Cosmatos *et al.*, 1976), but its occurrence detracts from the attractiveness of routes that involve saponification.

Weitzel *et al.* (1976) have reported fragment-condensation semisyntheses of analogues directed toward the study of the same region as those mentioned above. They, too, prepared N^{α}-Gly^{A1}, N^{α}-Phe^{B1}-di-Boc-des-octapeptide-(B23–B30)-insulin pentamethyl ester. They coupled it to the methyl esters of the appropriate materials to prepare, after deprotection, des-heptapeptide-(B24–B30)-insulin, des-hexapeptide-(B25–B30)-insulin, des-pentapeptide-(B26–B30)-insulin, des-tetrapeptide-(B27–B30)-insulin, [Ala^{B24}]des-hexapeptide-(B25–B30)-insulin, [Ala^{B25}]des-pentapeptide-(B26–B30)-insulin, [Ala^{B26}]des-tetrapeptide-(B27–B30)-insulin, and [Ala^{B27}]des-tripeptide-(B28–B30)-insulin. Weitzel *et al.* describe the many experiments that they carried out to establish the best conditions for each step, and discuss the ways in which Ruttenberg's conditions had to be modified in order to make the scheme work. They found that the couplings rarely went to completion, although the yields, usually between 50–80%, were not impossibly bad. They appear to have chosen a method of saponification designed to be as mild as possible consistent with efficiency. They report that insulin hexamethyl ester deprotected by their method has much the same activity as the native hormone in the mouse-fall test. An electropherogram suggests that there was no amide loss. No tests are described, such as crystallization or electrophoresis of proteolytic digests, that might have detected transpeptidization, should it have occurred in regions of the molecule that are not important for activity in the mouse-fall test. The authors report comprehensive, statistically significant, assays of their analogues by this test. The results enable them to draw unambiguous conclusions about the functional role of residues in the region B23–B27. They have also compared and contrasted the properties of their semisynthetic analogues with those of the earlier authors, and with those of their own, totally synthetic, analogues.

More recently, Weitzel *et al.* (1978) have used carboxypeptidase B to remove Arg^{B22} from des-octapeptide-(B23–B30)-insulin pentamethyl ester. After protection of the amino termini by the Boc- group, the des-nonapeptide-(B22–B30)-insulin ester was coupled to various tetrapeptide methyl esters of the sequence X-Gly-Phe-Phe-OMe. The coupled products were deprotected and purified. X, which represents position B22, was Arg, N^{ϵ}-Boc-Lys, N^{δ}-Boc-Orn, citrulline, Ala, or Gly. They measured the biological activity of these analogues and further confirmed the essential role in activity of Arg^{B22}, and of its guanido group.

Experience in this Laboratory over a number of years (H. T. Storey, A. R. Rees, and R. E. Offord, unpublished work) confirms much of what has been said about the reactions involved. We have wished to avoid the methyl group and have found ourselves obliged to use diazoalkanes. As a consequence we have had to protect the side-chains of the histidine residues against *N*-alkylation. The Boc—NH—$CH(CF_3)$- group is satisfactory. After trying several diazoalkanes, Offord *et al.* (1976) found that *p*-methoxyphenyldiazomethane was the best reagent to use for —COOH protection in insulin. The fully protected insulin gives a crystallizable product on deprotection. We, too, have found that couplings do not go to completion.

In summary, although the route proposed in 1969 appears to be particularly suitable for use on insulin, it has been found in practice to yield worthwhile results only with difficulty. We shall see that the route has been easier to apply in other protein systems, but Obermeier and Geiger (1976) and Obermeier (1978) have to some extent cut through the difficulties of its application to insulin. They simply reacted N^{α}-Gly^{A1}, N^{α}-Phe^{B1}-di-Boc-des-octapeptide-(B23–B30)-insulin (i.e. there was no side-chain protection) with dicyclohexylcarbodiimide and 1-hydroxybenzotriazole and a partially protected synthetic octapeptide. They were rewarded for this gesture of faith by a usable yield of the correct product, which could be separated from isomeric side-products by ion-exchange chromatography, and which did not need to be deprotected by saponification. The yield of the correct product (25%) suggests that this might be an instance of a sterically assisted coupling between minimally protected fragments, as discussed in other Chapters, and by Rees and Offord (1976a). Obermeier (1978) has used his route to produce 4 g of semisynthetic human insulin from porcine insulin.

By contrast, Gattner *et al.* (1978) found that the analogous route to human insulin *via* the product of limited peptic cleavage of porcine insulin, the des-pentapeptide-(B26–B30) derivative, was only 1 % efficient.

Obermeier (1978) has reported the use of the des-octapeptide route for the production of analogues. He has made des-Thr^{B30}-[Arg^{B29}]insulin, des-tetrapeptide-(B27–B30)-[des-carboxy-Tyr^{B26}]insulin, and, as a crude product, [D-Ala^{B23}]insulin.

Certain of the analogues of insulin dealt with under the heading of stepwise synthesis were, strictly speaking, the products of fragment condensations. Among these was the work of Shimonishi (1970). They were considered with the stepwise semisyntheses because the natural fragments were prepared for coupling by stepwise degradation, if at all, and because the synthetic peptides were usually very short.

Lysozyme

Hen-egg lysozyme has 129 residues, of which 11 are arginine, 6 are lysine and 6 tryptophan; there are 4 disulphide bridges (Canfield, 1963; Jollès *et al.*, 1963). N^{ϵ}-Acylation, therefore, still leaves 11 sites for tryptic cleavage. There is also the problem of the cleavage of the bridges, the protection of the cysteine residues, and the re-formation of the bridges at the end of the operation. These facts, and the relatively high content of tryptophan (of which, as we saw in earlier Chapters, the fewer the better), do not make lysozyme an attractive subject for fragment-condensation semisynthesis.

By the same token, however, the protein presents an excellent proving ground for the techniques: almost every conceivable problem is likely to be met with. Backer and Offord (1968), Offord (1969a), Hayward and Offord (1971), Offord (1973), Rees and Offord (1976a, b), and Offord *et al.* (1976) have reported semisynthetic techniques of general application that were developed on the tryptic peptides of lysozyme. Offord (1969a, 1973) and Rees and Offord (1976b) have coupled certain of the peptides together to produce fragments of between 31 and 51 residues in length (Fig. 8.2).

Those fragments drawn from the first 40 or so residues of lysozyme have some interest in their own right. The conformation of this region is stabilized to a striking extent by non-covalent interactions between its own residues, and very much less so by interactions between these residues and those in the remainder of the protein (Blake *et al.*, 1965). This impression of being self-contained is further enhanced by the stretches of repeating, hydrogen-bonded structure in the region, and the way in which the distribution of its hydrophobic and hydrophilic residues between the surface and the interior resembles that of an independent protein. Since protein biosynthesis begins from the amino terminus, it was suggested that these structural features may testify to the regions acting as a former against which the later portions of the sequence take up their conformation. Physical studies on those of the fragments in Fig. 8.2 that relate to the region in question do not support the hypothesis, though they do not positively disprove it (A. R. Rees, J. T. Thornton, and R. E. Offord, unpublished results).

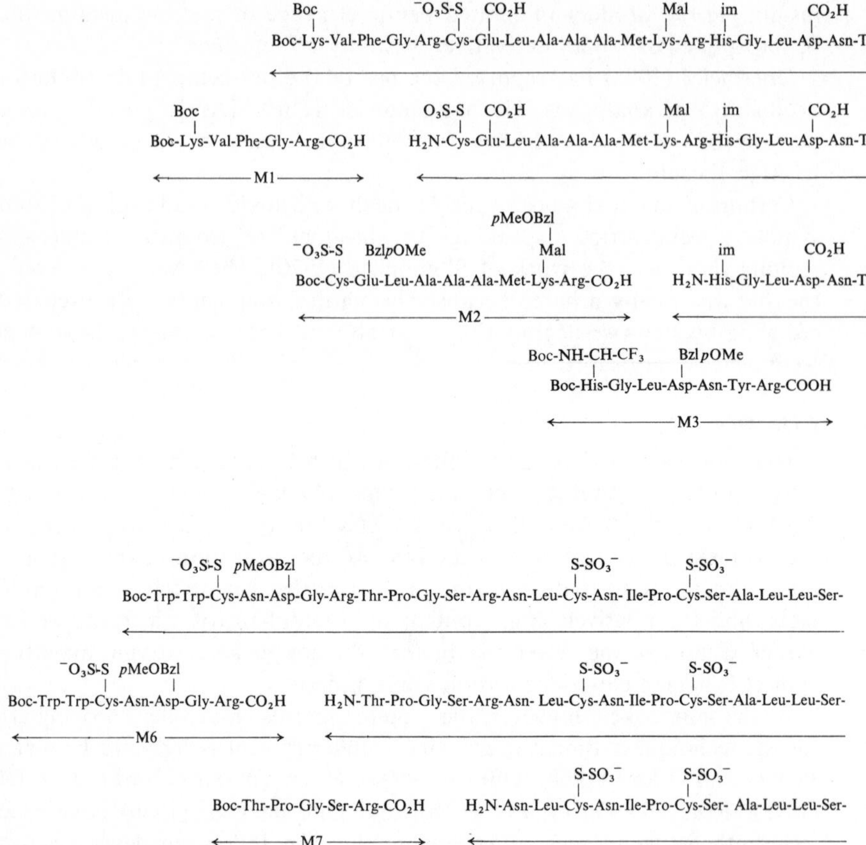

Fig. 8.2. Condensation of lysozyme fragments. The sequences of the fragments are taken from Canfield (1963) and Jollès *et al.* (1963). Differences between these two sources were reinvestigated by Rees and Offord (1972). The following abbreviations are used: Boc—NH—CH(CF_3)-, 1-t-butoxycarbonylamido-2,2,2-trifluoroethyl-; *p*MeO-Bzl-, *p*-methoxybenzyl-; Mal-, maleyl-; im, unprotected imidazole side-chain. [Reproduced from Rees and Offord (1976b) by permission of the Biochemical Society.]

Ferredoxin

Most of the semisynthetic work on the ferredoxins has been of the stepwise type and was considered under that heading. Lode *et al.* (1974b) have extended their studies on the role of aromatic residues at positions 2 and 30 (see p. 169) by carrying out fragment condensation. *Clostridium m-e* ferredoxin has tyrosine at position 2, but a non-aromatic residue, arginine, at position 30. Lode *et al.* cleaved the Boc- apo-protein with chymotrypsin and then coupled Boc-Ala-Leu to the N^{ϵ}-di-Boc-des-dipeptide-(1–2)-apoferredoxin. After deprotection and reconstitution with iron and sulphide, the [Leu^2] analogue was found to have the same activity as the natural protein, though the binding of the iron–sulphur clusters was somewhat less stable. It therefore appears to be confirmed

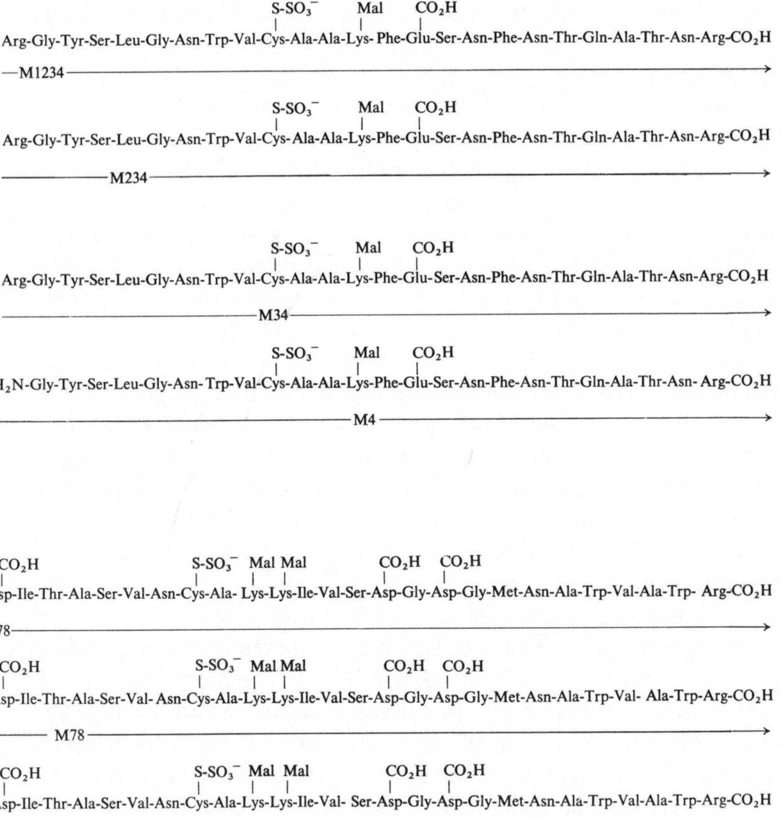

that the aromaticity of the residues in these positions is not essential for electron transport through the protein. The authors were fortunate that some of the non-aromatic residues that sometimes provide cleavage sites for chymotrypsin seem not to have done so on this occasion, although they are present in the protein. The authors draw attention to evidence of under-protection of the ε-NH_2 groups of the des-dipeptide ferredoxin (they used a markedly lower concentration of t-butyl azidoformate than in some of the Examples in this book), which gave rise to some spurious products. Nonetheless, the conclusions that they draw as to the relationship between structure and function are unlikely to be invalid.

Nucleases

Bovine pancreatic ribonuclease has 10 lysine residues, 4 arginine residues, 4 methionine residues, and 4 disulphide bridges (Smyth *et al.*, 1963). Apart from its lack of tryptophan, it is only a little more suitable for extensive semisynthetic work than is lysozyme (see above). Offord (1972) used the CNBr peptide corresponding to residues 1–13 for model studies on the use of the lactone form of homoserine for making a distinction between α- and side-chain —COOH groups.

Staphylococcus aureus nuclease lacks disulphide bridges but its composition does not otherwise make it much more favourable for semisynthetic work than the pancreatic enzyme. However, Di Bello (1975) has exploited the fact, mentioned earlier, that the products of limited tryptic digestion can associate to form a non-covalent complex with some enzymic activity. One of these fragments, corresponding to residues 6–48 (or 6–49), has only one arginine residue, at position 35. Di Bello has carried out some model studies directed to using semisynthetic techniques to speed up the production of analogues of this fragment of reliable structure. Recognizing that, for the reasons discussed in Chapter 1, the products of solid-phase syntheses become very much less reliable as their length increases, Di Bello used the solid-phase method to produce the sequence corresponding to residues 36–47. (Fragment 6–47 is as effective in forming an enzymatically active complex as is fragment 6–48 or 6–49.) He then condensed the synthetic peptide, while still on the solid phases with $N^{\alpha,\epsilon}$-trifluoroacetyl, O^{ω}-benzyl fragment 6–35, which he prepared by Offord's (1969a) scheme. Like Offord (1973) and Rees and Offord (1976b), he used dicyclohexylcarbodiimide and 1-hydroxybenzotriazole. A model coupling gave no evidence of racemization. Di Bello cleaved the product from the resin, with simultaneous removal of the benzyl groups, by means of HF/anisole. He removed the trifluoroacetyl groups with 1 M piperidine. The crude semisynthetic material was alread 2–3 times as active in the formation of a functioning complex as was totally synthetic fragment 6–47, but still noticeably less active than the natural peptide. Di Bello points out that incomplete removal of the trifluoroacetyl groups and side-reactions in the HF/anisole system could have been the cause of the loss of activity of the semisynthetic product. To these possible causes one could add the fact that, since this was a preliminary study, Di Bello did not protect histidine from the action of the diazoalkane. Histidine is not involved in the catalytic activity of the enzyme, but the effect of alkylation could be to impair the formation of non-covalent complex. In a full study, it would probably be desirable to protect histidine, and either to try a more labile —NH_2-protecting agent or to test one of the milder procedures for removal of the trifluoroacetyl group (e.g. Example (7.3)). The availability of the *p*-methoxybenzyl group for semisynthesis (Offord *et al.*, 1976) means that very strong acids, such as HF, are no longer needed for such work.

The fragment possesses two methionine residues, at positions 26 and 32. This offers the opportunity of extending Di Bello's approach to other parts of the fragment and in some instances being able to avoid conventional synthesis altogether.

In contrast to the two nucleases mentioned, ribonuclease-T1 is very tempting for general semisynthetic work. Although it has two disulphide bridges, only one of its 104 residues is arginine, one is lysine and there are no methionine residues (Takahashi, 1965). The lysine is at position 41 and the arginine is at position 77, which would permit access to two separate regions of the protein.

The absence of methionine means that the protecting groups labile to CNBr (e.g. Saunders and Offord, 1977a, b) could be used if necessary.

Cytochrome c

The spontaneous re-formation of the bond 65–66 when the appropriate CNBr fragments of horse-heart cytochrome *c* are mixed under reducing conditions has been mentioned several times before. The fragments are those corresponding to residues 1–65 and 66–104 and the product of the spontaneous coupling is itself an analogue, [homoserine-65] cytochrome *c* (Corradin and Harbury, 1971, 1974b; Harbury, 1978). This product is nearly indistinguishable in its properties from the natural protein: serine, among several other residues, appears at this position in the cytochromes *c* of lower organisms (Dayhoff, 1972). Corradin and Harbury (1974a) have produced hybrid molecules by allowing peptides from the beef and tuna-fish cytochromes to combine. Wilgus and Stellwagen (1974) have combined N^{ϵ}-guanidinated peptide 66–104 with unmodified peptide 1–65 and studied the properties of the resulting partially guanidinated protein.

Harbury (1978) has reviewed these and more recent developments at length. In particular, he discusses experiments by P. Berman and himself on the major antigenic determinant of cytochrome *c*. Hybrids were made between fragments of some closely related cytochromes: horse (1–65) : cow (66–104); cow (1–65) : horse (66–104); and rabbit (1–65) : cow (66–104). If we ignore the conversion of residue 65 to homoserine, the first two hybrids are in effect the [Gly^{89}] and the [Ser^{47},Gly^{60}] analogues of the horse protein, while the third is the [Val^{44},Asp^{62}] analogue of the rabbit protein. The antibody-binding behaviour of these hydrids was studied, together with that of the natural protein from horse, cow, rabbit, and donkey. The results established that the residue responsible for the difference in binding between the cytochromes *c* of horse and cow is at position 60, and that responsible for the difference between rabbit and cow is at position 44.

Corradin and Chiller (1978) have used the horse: cow and rabbit: cow hybrids to study the specificity of T-lymphocyte activation in mice.

Harbury's review describes a number of 1 : 1 complexes, some of which have biological activity, in which the covalent coupling does not occur because the fragments overlap or because they are too short. Among these are included hybrids made up of fragments of cytochrome from horse and the yeast *Candida*, and of horse and the bacterium *R. rubrium*. Harbury also describes the total synthesis of residues 66–104 of the horse sequence, in preparation for the covalent semisynthesis of analogues of the whole, 1–104 residue molecule. Related, but technically very dissimilar, studies directed to the same end are being carried out by Boon *et al.* (1978) and by Warme and coworkers (Ledden *et al.*, 1977; Kirby and Warme, 1978).

Wallace (1976) has adopted the Corradin and Harbury coupling to N^{ϵ}-acetimidylated fragments. He was also able to find ways of coupling under oxidizing conditions, though with lower efficiency than under reducing conditions.

Wallace (1978) and Wallace and Offord (1979) have produced the peptide 66–104 by fragment condensation and then combined it with the peptide 1–66 by Corradin and Harbury's method. Scheme (8.8) illustrates the different approaches to the production of analogues of cytochrome *c* that are currently being employed.

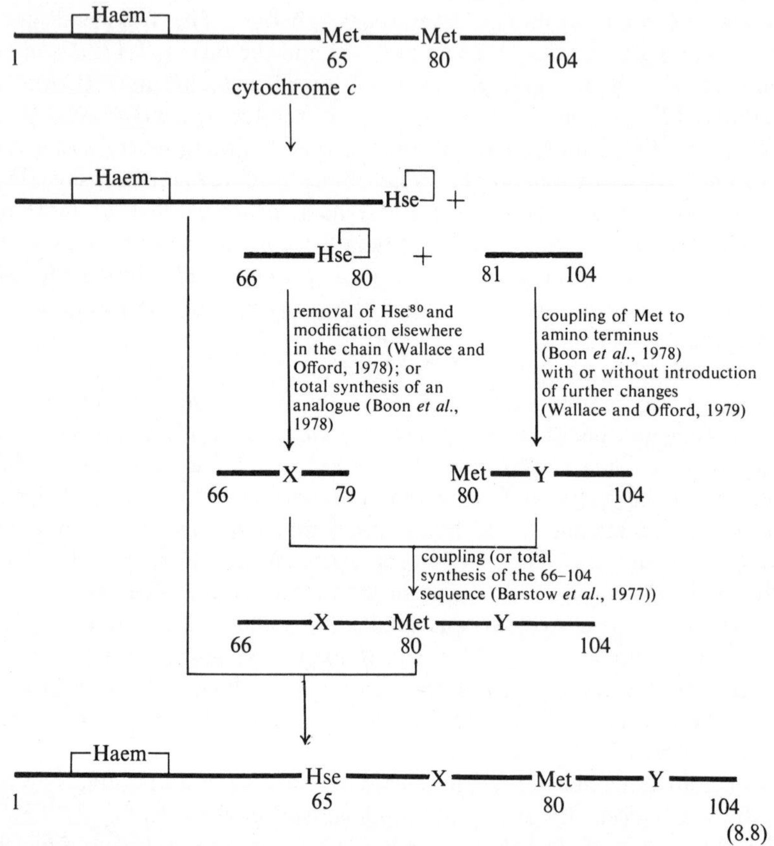

(8.8)

Wallace (1976, and unpublished experiments with R. E. Offord) used the azide coupling scheme of Offord (1972) (see p. 122) to couple the two peptides, 1–65 and 66–104, before the spontaneous coupling was discovered. Yields were poor and, apart from the natural inefficiency of the azide process, this may have been true in part to the strong tendency (observed in model experiments) of homoserine azide to revert to the lactone. Indeed, such coupling as there was may have been via the re-formed lactone. A further difficulty is that residue 66, a glutamic acid (Margoliash *et al.*, 1961), when amino terminal, tends rather readily to form the pyrrolidone carboxylic acid on the type of handling received during the azide coupling. This side-reaction would naturally lower the coupling efficiency. Wallace (1978) dealt with the problem by making the [O^{β}-t-butyl-Asp^{66}] analogue of N^{ϵ}-acetimidyl-peptide 66–104 by stepwise methods. He also found that methylation of the —COOH groups of the peptide with methanol/HCl prevented the side-reaction from occurring.

Harris (1977) and Wallace and Offord (1979) have examined the protection of cytochrome *c* and its fragments by diazoalkanes and by methanol/HCl. They have also studied the effect of subsequent deprotection. The *p*-methoxybenzyl group is satisfactory for protection of the CNBr fragment 66–80, although it is unusually base labile when on Glu^{69}. The group is less suitable for the tryptic haem peptide, 1–38, in which both the peptide and the haem —COOH groups are protected. Even when the acetimidyl group is used for N^{ϵ}-protection the esterified peptide is very insoluble in the sort of solvents required for enzymic deprotection of the α-COOH and for characterization of the product. The position is made worse if the side-chains of the histidine are protected against alkylation by the diazoalkanes. There also appears to be a marked loss of colour from the haem peptide. Methanol/HCl produces a much more tractable product when applied to this peptide. We find no evidence that transpeptidation (Reaction (1.20)) is a serious drawback on deprotection of the methyl ester of intact cytochrome *c*. There is no loss of biological activity on protection and subsequent deprotection, no change in spectroscopic properties, and no change in the pattern on electrophoresis of proteolytically digested material. Cytochrome *c* might be an exceptionally favourable candidate for methyl protection because it may well have a tight configuration even under the conditions of saponification (pH 10). This would tend to restrict the participation of the peptide backbone in the transpeptidation mechanism. Other proteins, and perhaps even some fragments of cytochrome *c* itself, might not give such a satisfactory outcome.

The protein is an excellent subject for semisynthetic techniques in general. Many analogues that are required for the elucidation of the means by which the electrons pass through the protein, and the mode of interaction between cytochrome *c* and other mitochondrial components, can only be made semisynthetically, rather than by exploiting side-chain reactivity of the intact protein. Conventional synthesis of the molecule appears to be at least as difficult as its chain length would lead one to expect, and in addition one has to overcome the hurdle of fastening the haem ring to the polypeptide chain.

Myoglobin

In addition to the non-covalent and stepwise semisyntheses already mentioned, three groups of workers have reported steps toward fragment-condensation work. Hagenmeier *et al.* (1978) have coupled the partially protected fragments 1–31 and 32–139 of horse myoglobin. Although the amino-acid composition of the product was satisfactory, in this preliminary study the same, very severe conditions were used as for the work on the fragment of staphylococcal nuclease discussed on p. 182. One would expect the product to be damaged in the various ways already described and indeed Hagenmeier *et al.* (1978) report that it had only 12% of the theoretical ability to form the correct, myoglobin-like complex with haem. Nonetheless the work shows that the general scheme for fragment condensation will produce products of substantial chain length, and in view of the existence of milder alternatives for the protection and deprotection steps, the results must be regarded as extremely encouraging.

Borrás *et al.* (1978) have developed methods for the isolation in a very good state of purity of the three tryptic fragments of $N^{\alpha,\epsilon}$-maleyl horse myoglobin.

Wang *et al.* (1978) have produced the full 153-residue sequence of whale myoglobin by fragment condensation, as a pilot study for the production of analogues. They protected the ε-NH_2 groups by acetimidylation and cleaved at Trp^{14} by 2-(2-nitrophenylsulphenyl)-3-bromo-3′-methylindolenine (BNPS-skatole). They then reacted N^{α}-Boc-, N^{im}-formyltryptophan hydroxysuccinimido ester with the 15–153 fragment in order to replace the tryptophan. A synthetic peptide, corresponding to residues 6–13, was then condensed with the resulting semisynthetic 14–153 fragment. Finally, the complete 153-residue sequence was reconstituted by coupling on a peptide corresponding to residues 1–5. The couplings were followed by amino-acid analogues and amino-terminal determination. The reconstituted protein was totally deprotected and, after purification, was found to give satisfactory results when examined by absorption spectroscopy, circular dichroism, and electrophoresis. The prospects for the fairly rapid production of useful analogues seem excellent.

Carbonic anhydrase B

Human carbonic anhydrase B has 264 residues, of which 7 are arginine, 19 are lysine, 2 are methionine, 6 are tryptophan, and 1 is cysteine (Andersson *et al.*, 1972; Henderson *et al.*, 1973). In spite of the potential problems implied by this composition, a start has been made towards fragment-condensation semisynthesis of the enzyme (Fölsch, 1974, 1975, 1978). Fölsch carried out a detailed study of the protection of the 28 —COOH groups in the protein by phenyldiazomethane (benzyl esterification) and the water-soluble reagent triethyloxonium fluoroborate (ethyl esterification). He has derived optimum conditions for benzylation and found, as usual, that all residues other than hisitidine survived the reaction. Fölsch suggests that the dinitrophenyl group be applied to the histidines after —NH_2 protection but before esterification—although it should be remembered that the most common dinitrophenylating reagent, 3,5-dinitrofluorobenzene, also tends to react with the —OH groups of tyrosine. The resulting dinitrophenyl-ether bond is very hard to cleave, and consideration might profitably be given to the Boc—NH—CH(CF_3)- and Cbz—NH—CH(CF_3)- groups as alternatives.

Fölsch's studies on triethyloxonium fluoroborate are of particular interest. It might have been hoped that such a reagent would be free of the *N*-alkylating tendency of diazoalkanes, but in practice it proved even more prone to promote side-reactions. Even under conditions in which the —COOH groups are only partly esterified there is considerable loss of lysine (there was simultaneous production of a new, basic amino acid) and even some loss of tyrosine.

Triosephosphate isomerase

This protein has 247 residues, of which 8 are arginine, 2 are methionine, 4 cysteine (not in bridges), and 4 are tryptophan (Furth *et al.*, 1974). Milman (1974) and P. C. Mahony and R. E. Offord (unpublished work) have begun to explore the preparation and protection of CNBr fragments of the $N^{\alpha,\epsilon}$-acetimidyl enzyme. There are two reasons for wishing to prepare analogues near the CNBr cleavage points. Both methionine residues are involved in the association of the 247-residue subunit into the dimeric form in which the enzyme is normally found and one, Met^{13}, is close to the active site (Banner *et al.*, 1975).

Melanotropins and ACTH

What has so far found to be the longest of the melanotropin group of hormones, human β-MSH, is only 22 residues long (Harris, 1959). They therefore cannot be called proteins, but are mentioned here because they were used by Burton and Lande (1970), Lande and Burton (1971), and Lande (1971) as model systems for the study of fragment condensation.

Some of the operations were carried out without covalent protection of ε-NH_2 groups. It was hoped that the difference in pK between α- and ε-NH_2 groups could be exploited to direct acylation away from the ε-NH_2 groups by pH control. In the event, although semisynthetic [Lys^{10}]porcine β-MSH was prepared, yields were low and side-products did form through the ε-NH_2 groups. Indeed the careful experiments on reaction rates that were done in order to optimize the yield of authentic product were, paradoxically, of use in obtaining specific N^{ϵ}-acylation (Lande, 1971). This is not the only instance in which investigations on side-chain reactivity, carried out to aid a semisynthesis, have contributed to our knowledge of the means of specific side-chain modification of polypeptides and proteins.

Lande and Burton (1971) used the general scheme of Fig. 8.2 to prepare hybrids of α- and β-MSH (Reaction (8.9)). The products of the couplings contained many side-products as well as the wanted ones, possibly because of incomplete side-chain protection. Burton and Lande used diphenyldiazomethane for —COOH protection under conditions that might not have been sufficient to give an adequate degree of protection.

Van Nispen *et al.* (1977) used side-chain protection to control the coupling between tryptic fragments of α-MSH. They modified one of the two fragments by stepwise semisynthesis and their final product was the [Phe^9] analogue of the hormone. The work was notable for the use of the methylsuphonylethyloxycarbonyl (Msc-) group for N^{ϵ}-protection and the t-butyl group for side-chain —COOH protection. Boon *et al.* (1978) further demonstrated the value of the Msc- group in semisynthesis by cleaving the $N^{\alpha,\epsilon}$-Msc- derivative of residues 1–24 of ACTH with CNBr at residue 4. They then coupled back the appropriate synthetic N^{α}-Boc-tetrapeptide onto the fragment 5–24. After treatment of the coupled product with trifluoroacetic acid only the α-NH_2 group was liberated.

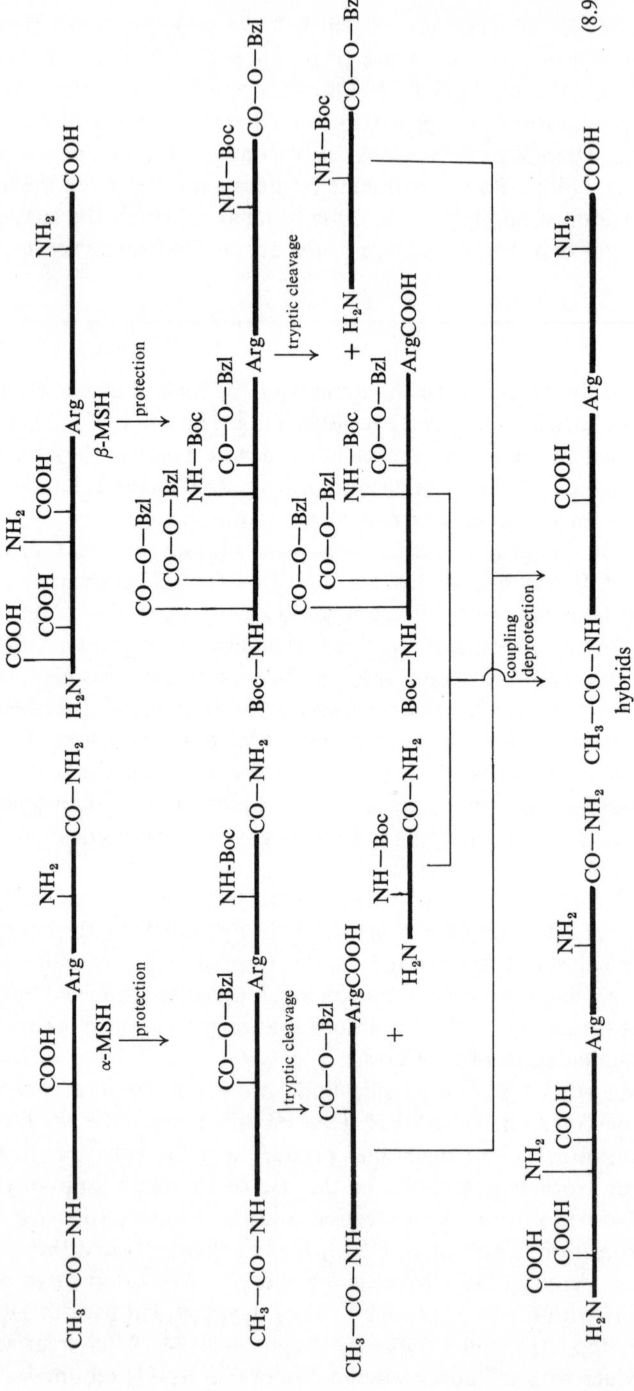

CH$_3$—CO—NH
COOH
Arg
NH$_2$
CO—NH$_2$
α-MSH
protection
CH$_3$—CO—NH
CO—O—Bzl
Arg
NH-Boc
CO—NH$_2$
tryptic cleavage
CH$_3$—CO—NH
CO—O—Bzl
ArgCOOH
+
H$_2$N
NH—Boc
CO—NH$_2$
H$_2$N
COOH
COOH
NH$_2$
COOH
Arg
NH$_2$
COOH
β-MSH
protection
Boc—NH
CO—O—Bzl
CO—O—Bzl
NH—Boc
CO—O—Bzl
Arg
NH—Boc
CO—O—Bzl
tryptic cleavage
Boc—NH
CO—O—Bzl
CO—O—Bzl
NH—Boc
CO—O—Bzl
ArgCOOH
+ H$_2$N
NH—Boc
CO—O—Bzl
coupling
deprotection
H$_2$N
COOH
COOH
NH$_2$
COOH
Arg
NH$_2$
CO—NH$_2$
CH$_3$—CO—NH
COOH
Arg
NH$_2$
COOH
hybrids

(8.9)

The resulting N^{ϵ} fragment of ACTH was found to have unimpaired lipotropic activity.

Immunoglobulins

A brief report has appeared (Hancock and Adams, 1974) of studies toward the semisynthesis of immunoglobulins. It appears that the scheme of Fig. 8.2 is intended, rather than the use of disulphide bridges or the non-covalent approach used by Gavish *et al.* (1978), to produce hybrid antibodies. If so, in view of the size, heterogeneity, and number of points of cleavage of these molecules one can only applaud a most courageous start.

Clearly, the most likely applications would exploit the existence of constant and variable regions in the antibody structure. One could envisage the coupling of a synthetic variable region to a natural constant region, which would greatly reduce the number of steps needed to prepare antibody of the desired sequence. Since the total chemical synthesis of an antibody has been contemplated (see Haber and Krause, 1977), it seems worthwhile to make these remarks, but it seems very likely that selective cell-culture methods will prove a far more profitable line of approach than any chemical operations.

Protease inhibitors

Soyabean trypsin inhibitor (Kunitz type) has 181 residues of which 9 are arginine, 10 are lysine, 2 methionine, and 4 tryptophan; there are 2 disulphide bridges (Koide *et al.*, 1972). In spite of the size of the protein, important semisynthetic modifications have been made by exploiting the mechanism that underlies the protein's biological activity.

The active site of the inhibitor is the peptide bond between Arg^{63} and Ile^{64}. A catalytic quantity of trypsin (1–2 mol per cent) cleaves the bond without having an effect on inhibitory activity. Both the untreated inhibitor and that modified by the cleavage form a 1 : 1 complex with trypsin and depend for their inhibitory properties on the great strength of this association ($K_i = 10^{-8}$ M). Sealock and Laskowski (1969) made use of the fact that the complex between the modified inhibitor and trypsin can be dissociated by acid or denaturing solvents with simultaneous and almost complete re-formation of the Arg-Ile bond.* They used carboxypeptidase B to remove the arginine and to catalyse its replacement with lysine. The replacement reaction would normally only go to about 1% of completion, but the equilibrium is pulled over by the very much stronger binding capacity of trypsin for the modified inhibitor (or in this instance for its [Lys^{63}] analogue) then it has for the des-Arg^{63} form of the modified

* This might at first sight be thought to violate the principle of microscopic reversibility, since an enzyme (trypsin) is involved in altering the position of equilibrium (the balance between the modified and intact forms of the inhibitor) rather than the rate of attainment of that equilibrium. Sealock and Laskowski point out that trypsin is present in reactant rather than catalytic quantities and show that, as a consequence, no thermodynamic principles are violated. If the interaction between the enzyme and the inhibitor is such as to be able to contrive the re-formation of the bond when they dissociate, it may do so.

inhibitor. The series of reactions is closed by the dissociation of the analogue–trypsin complex to form the [Lys^{63}] inhibitor (Reaction (8.10)).

Leary and Laskowski (1973) used the same principle to produce [Trp^{63}] inhibitor. Carboxypeptidase A was used to catalyse the coupling of tryptophan to the des-Arg^{63}-modified inhibitor, which had been produced by carboxypeptidase B, as before. The equilibrium governing the coupling of tryptophan

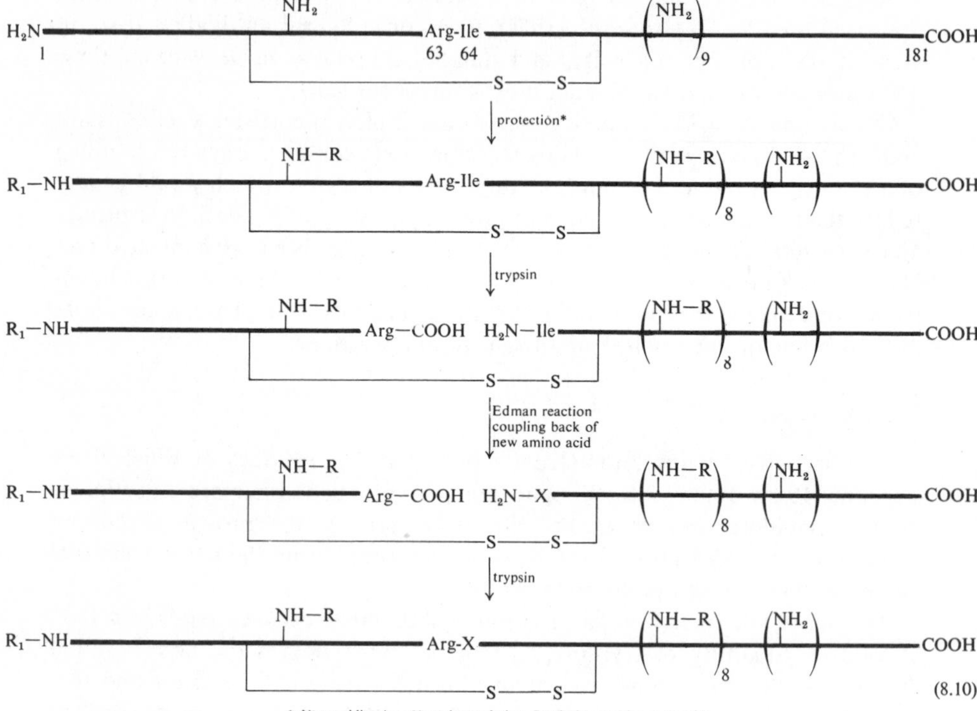

* N^{ϵ}-guanidination, N^{α}-carbamoylation. One lysine remains unreactive.

was pulled over in the direction of synthesis by permitting a complex to form between the product and chymotrypsin, rather than trypsin as before. The specificity site of chymotrypsin, as is well known, binds such amino acids as tryptophan with the same sort of intensity as the specificity site of trypsin shows for arginine. The [Trp^{63}] analogue bound very strongly to chymotrypsin and weakly, if at all, to trypsin. The authors do not comment on the fact that Tyr^{62} ought in principle also to be susceptible to carboxypeptidase A, and therefore might have become involved in this sequence of reactions.

This experiment is of particular interest in that it represents the production, by design, of a new biological specificity.

More recently, Kowalski and Laskowski (1976a) have combined chemical operations with this purely enzymatic approach. They first protected the ε-NH_2 groups by guanidination (which has the advantages, except for reversibility, of acetimidylation) and opened the bond 63–64 in the usual way.

They then removed Ile^{64} by the Edman reaction and replaced the residue by active-ester couplings to make the [Ala^{64}] and [Leu^{64}] analogues of the modified inhibitor. They also reinserted isoleucine in position 64. The [Ala^{64}] analogue of the intact inhibitor was formed by dissociating the trypsin–analogue complex.

Kowalski and Laskowski (1976b) have used analogous methods to carry out chain extension on Ile^{64} of the modified inhibitor. They used active esters to add alanine and isoleucine to the chain. They also used the *N*-carboxyanhydride method to add glutamic acid in this position. Trypsin (5 mol per cent) was used to catalyse the formation of the peptide bond. These analogues would not form the 1 : 1 complex and so the customary route to the re-formation of the intact inhibitor was not available. The bond in the Ala and Leu analogues re-formed only to a small extent, while it did so in the Glu analogue to the extent of 16 %. Further interpretation of the data obtained from studies on these analogues has been given by Kowalski (1978) including the correlation of the activities of some semisynthetic analogues with X-ray crystallographic data on the enzyme–inhibitor complex.

Bovine trypsin-kallikrein inhibitor (Kunitz type; synonyms include: basic pancreatic trypsin inhibitor and Trasylol—the latter is a registered trade mark of Bayer A.G.) has 58 residues of which 6 are arginine, 4 are lysine, 1 is methionine; there are 3 disulphide bridges (Kassell and Laskowski, 1965). Although the composition does not make it seem especially attractive for semisynthetic work, the findings of Dyckes *et al.* (1974, 1977, 1978) concerning the effect on the protein of cleavage with CNBr have altered the position considerably. The one methionine residue, Met^{52}, lies between two disulphide bridges and the two fragments do not, therefore, separate from one another after cleavage. Dyckes *et al.* found that the broken bond spontaneously re-formed in neutral solution in the course of a few days, with restoration of biological activity. The

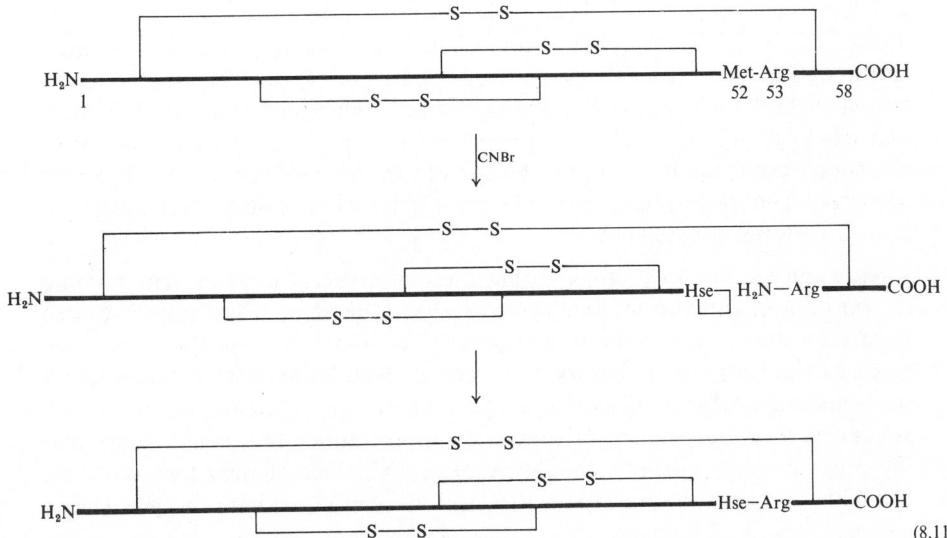

resulting analogue is the [homoserine-52] protein (Scheme 8.11). This analogue has been used, by comparison of its behaviour with that of the natural protein and a truncated derivative, in a detailed study on the mechanism of refolding after denaturation (Creighton *et al.*, 1978).

The discovery of this coupling is similar to the observation of Corradin and Harbury (1974a) of spontaneous coupling in cytochrome *c*. It seems probable that these couplings are sterically assisted in the way discussed on p. 29 and, indeed, Dyckes *et al.* (1977) have shown that the rate of the coupling drops sharply in denaturing solvents (aqueous guanidinium chloride and dimethylsulphoxide). This property of denaturing solvents opened the way to the preparation of other analogues. Without denaturants, the natural hexapeptide (Scheme 8.11) pushes aside any synthetic peptide that is intended to replace it even if the synthetic molecule is present in much higher concentrations. In the presence of denaturants, the relative rates of the intra- and intermolecular couplings more nearly reflect the relative concentrations of the reactants. Once a synthetic peptide has been incorporated, the natural one can be detached from the analogue by reduction of all the disulphide bridges. By this means Dyckes *et al.* (1977) coupled synthetic Gly-Gly-Ala to residue 52 and made the reduced form of homoserine-52-des-tripeptide-(53–56) inhibitor (Scheme 8.11). Attempts to reoxidize this analogue to form the correct disulphide bridge appeared to be wholly unsuccessful, but the reduced form itself had 45% of the molar inhibitory capacity of the native inhibitor.

Jering and Tschesche (1974, 1976a, b, 1978) have used a similar series of reactions to that of Laskowski and coworkers to prepare analogues of the trypsin-kallikrein inhibitor. The crucial step was to isolate the modified (cleaved-chain) form of the protein—in this case the bond is between Lys^{15} and Ala^{16}. To do so proved more difficult than it had with the soyabean inhibitor: the problem was solved by weakening the tertiary structure by the selective reduction of one of the disulphide bridges. The bridge was re-formed by oxidation at the end of the sequence of reactions. Exopeptidases were used to make des-Ala^{16}-Arg^{17}-Ile^{18} inhibitor, des-Ala^{16}-Arg^{17} inhibitor, des-Ala^{16} inhibitor, and des-Lys^{15} inhibitor. [Arg^{15}], [Phe^{15}], and [Trp^{15}] analogues of the intact inhibitor were made by similar techniques to those described for the soyabean inhibitor. The chain-intact forms of these analogues showed previously unknown patterns of specificity.

Although we saw in Chapter 6 that there are several ways of manipulating the position of equilibrium in enzyme-catalysed coupling, it is always better to have substrates in which the fragments are held together in strong association (such as the protease inhibitors, but perhaps also some of the non-covalent semisynthetic analogues discussed earlier). The position of equilibrium in such systems is then independent of substrate concentration. This is not to say that the other possible methods should be neglected, since, if only they could be brought into general application, they would greatly facilitate the production of analogues of all kinds.

Note added in proof. Probably the most important development that has taken place in the period during which this book has been in the press is the great increase in the number of successful couplings by the method of enzymic catalysis. Two particularly important papers on the use of trypsin in the enzymic transformation of porcine into human insulin have appeared. One is by K. Morihara, T. Oka and H. Tsuzuki (Nature **280**, 412–413 (1979)) and the other is by K. Inouye, K. Watanabe, K. Morihara, T. Tochino, T. Kanaya, J. Emura and S. Sakakibara (J. Amer. Chem. Soc. **101**, 751–752 (1979)). The enzyme clostripain has been used to join the fragments in non-covalent complexes derived from ribonuclease A, Staphylococcal nuclease and (in preliminary experiments) cytochrome *c* (G. A. Homandberg, A. Komoriya, M. Juillerat and I. M. Chaiken in 'Proceedings of the Sixth American Peptide Symposium' edited by E. Gross (in Press)).

CONCLUSION

Although this book has from time to time pointed to the advantages of semisynthesis over total synthesis in particular applications, this is not to suggest that the semisynthetic approach is always simple to apply, or that the structural integrity of its products is always beyond question. However, the overall impression one receives is that, although there is a need for a continuing improvement in the method, it can be, and is being, applied with reasonable confidence to the solution of problems that could previously be tackled, if at all, only by total synthesis.

APPENDIX 1

Characterization of products and intermediates

Synthesis is normally seen in organic chemistry as the final check on the correctness of structural analysis. There is no doubt that the position is reversed in the case of proteins. When one is dealing with a small molecule it is safe to assume that identity in behaviour of synthetic and natural product is a fair indication of the identity of their structure. In a molecule which consists of thousands of atoms the misplacement of one or two might alter its physical, biological, or immunological properties to a measurable extent, but it equally well might not. Conversely, the failure of a synthetic sample to match a natural one in its properties is not a safe ground for rejecting the proposed structure on which the synthesis was based. Perturbations in the tertiary structure might be the cause of the difference, even if the covalent structure was correct. In any case the methods of sequence analysis are easier than those of synthesis and their results, though still subject to error, more trustworthy.

The implication seems to be that the authenticity of semisynthetic products must be verified by a full sequence analysis, but this requirement would rarely be acceptable in practice. (The case for a full sequence determination is as strong, if not stronger, for products of conventional synthesis but, understandably, also seems not to have been persuasive.) It is probably enough to settle for a more limited chemical characterization of the type that can be carried out, quickly, on a few tens of nmol of material. We should begin with the determination of the amino-acid composition (Example (A.1)) and the identification, and if need be quantitation, of the amino-terminal group (Examples (A.2) and (A.3)).*

When we check on the products of a conventional synthesis, all residues are suspect to some extent. In a semisynthesis, the principal places at which to look for errors are at the bonds formed by the couplings and at the more reactive or more labile, side-chains. In this connection we can gain a great deal of information by examining a few of the electrophoretic properties of the proteolytic digests, in ways that we shall now discuss.

* Some of the techniques described in this Appendix might seem to be too well known to require description. I have included them mainly because workers in some countries may not have full access to the standard literature.

To be engaged on semisynthetic work on a protein more or less assumes that an amino-acid sequence has been determined. The experimental section of the paper describing the determination may provide valuable information on points of cleavage and patterns of peptide separation. For instance, it is usually possible to look up a method for cleaving and isolating a short fragment that contains the residues at which a coupling took place. The successful isolation of this bridge peptide is an assurance that coupling took place as planned. As sequence workers are turning more and more to automated methods, and in any case are tending to publish less experimental detail, such specific guide-lines may be more difficult to obtain in the future. But there still remains a number of general methods for examining our product. For instance, if the semisynthesis was carried out from tryptic fragments, the ability of trypsin to digest the product back to the original peptides (Example (A.4)) is a sign that racemization has probably been avoided. Although some protease will, contrary to expectation, cleave at D-residues, trypsin has yet to be shown to do so. (There are specific stains for some of the more reactive residues which can usefully be applied to the peptides to see if the side-chains have suffered damage—see Note 6 of the Example.) If the peptides are compared electrophoretically with those from a digest of the natural molecule, the extent of amide loss can be assessed. Example (A.5) gives the details of how quantitative estimates can be made of the number of amide groups per peptide. The Example also explains how one looks for evidence of transpeptidation between α- and side-chain —COOH groups (pp. 15–16, 138). The electrophoretic method is equally applicable to assessing the extent of $—NH_2$ or —COOH protection or deprotection (Example (A.6)), provided that the molecules are sufficiently soluble in the buffers used for the electrophoresis. The running properties of the larger and more heavily protected molecules can sometimes be improved by making the buffers 7–8 M in urea and using cellulose acetate rather than paper as the support (Example (A.7)).

All these electrophoretic methods are applicable in principle to the intact protein as well as to fragments. Isoelectric focussing in acrylamide gels in the presence of urea (Wrigley, 1968) is another method, most suited to proteins, for the estimation of net charge and thus state of protection or amidation (Example (A.8)).

Chromatographic, as well as electrophoretic, methods can be used. Although thin-layer chromatography is most readily applicable to small molecules (Example (A.9)), both it and paper chromatography (Example (A.10)) can sometimes be of service even with proteins. Gas–liquid chromatography, although it is rapid, sensitive, and of high resolving power, is not at first sight likely to have much application to this kind of work because only the smallest molecules are sufficiently volatile, even after chemical masking of all the polar functional groups of the main- and side-chains. But if a fragment of 20–30 residues (or less) can be digested with a fairly non-specific protease, the polar groups of the digest can be masked and a highly specific pattern of g.l.c. peaks observed. Better still, the effluent from the column can be led directly into a mass spectrophotometer, and very precise structural information obtained (Rose and Priddle, 1978).

However immaculate the results of the check on the covalent structure, they serve only as a prelude to the further evaluation of the product as a functioning entity. Each product will pose its own problems of investigation, involving the assay of biological and immunological activity, and examination by spectroscopic and other physical methods. Such an evaluation shades imperceptibly into the use of the material for research purposes. This Appendix ends with a number of figures that illustrate this gradation, and which also serve to emphasize the wide range of uses to which semisynthetic products can be put.

EXAMPLES

Example (A.1). The determination of amino-acid composition

Source. Standard methods, reviewed in Hirs (1967).

Process involved. Acid hydrolysis followed by automated ion-exchange chromatography on sulphonated polystyrene.

Method

The sample was transferred to a Pyrex tube (12 mm × 75 cm or 4 mm × 25 mm, as necessary). After drying down—if necessary—6 M HCl was added (100–200 times the weight of sample) and the tube sealed *in vacuo*. The tube was placed in the oven at 108 °C for 24 h, opened, and the hydrolysate dried down for analysis on an automated analyser.

Notes

(1) The HCl can be prepared as the constant-boiling fraction obtained by the distillation of conc. (12 M) HCl/water (1 : 1, v/v). Alternatively, if the conc. acid is of the highest available grade (e.g. British Drug Houses Aristar grade) it may be diluted and used without distillation. The acid should be stored in the dark, in the cold.

(2) The principal reason for preferring pure HCl is to avoid oxidation and chlorination of aromatic residues in the sample. It is possible to lessen the danger still further by adding phenol to the HCl to a concentration of 1% (w/v).

(3) In this Laboratory, the tubes are necked with an O_2/gas flame after the addition of acid. The tube is then connected to an arrangement that permits, alternately, exhaustion to an oil-pump vacuum and purging with O_2-free N_2. At least two cycles are employed and the neck is closed in the flame after a final

evacuation of the tube. Frothing of the sample can be controlled by freezing the acid in a small bath of acetone/solid CO_2. Safety dictates that large cooling baths should use isopropanol/solid CO_2, but this mixture does not cool the tubes quite well enough for the inexperienced operator to be sure of controlling the frothing.

(4) The Manufacturers' instructions will provide the best guide for the use of the analyser. It would not be helpful to give, for any particular system, elution volumes of the unusual amino acids likely to be encountered in semi-synthesis but the following data may provide a useful guide. Homoserine usually elutes after serine and before glutamic acid. When using some analyser resins, it is necessary to move its position of elution slightly by draining the hot-water jacket of the column for a few minutes after the loading of the sample. Homoserine lactone usually emerges near the position of histidine. It is usually converted to the open form before analysis by dissolving the dried hydrolysate in pyridine/acetic acid/water (25: 1: 225 by vol., pH 6.5), resealing the tube, heating to 108° C for 2 h, and drying down once more (Ambler, 1965). N^{ε}-Acetimidyl lysine usually emerges just after the NH_3 peak and well before arginine. It slowly forms lysine on hydrolysis (the apparent first-order rate constant at 108 °C is in the region of 0.0064 s^{-1} but depends so critically on temperature (cf. Reynolds, 1968, with Plapp and Kim, 1974) that it should be determined in the oven that it is intended to use for the hydrolysis). The *o*-, *m*- and *p*-fluorophenylalanines emerge separately, in the region of tyrosine and phenylalanine. The relative positions of emergence of many compounds can be deduced from the illustrations in Leggett Bailey (1967).

Example (A.2). End-group determination

Source. Based on Gray (1967, 1972).

Process involved.

H_2N—R_1 R_2 R_3 R_4···
peptide or protein
+
$N(CH_3)_2$
SO_2Cl
dansyl chloride

→

$N(CH_3)_2$
SO_2—NH—R_1 R_2 R_3 R_4···
dansyl peptide or protein,
(side-chains of Tyr and Lys
also react)

hydrolysis →

$N(CH_3)_2$
SO_2—NH—R_1—COOH

+ free amino acids,
side-chain substituted
Tyr and Lys

(A.1)

Method

A. *Peptides.* The peptide (about 10 nmol) was dissolved in 15 μl of 0.02 M $NaHCO_3$ and dried down *in vacuo* in a Durham tube. The dried material was redissolved in 15 μl of water and the pH checked by spotting a small amount onto pH paper. 15 μl of an acetone solution of 1-dimethylaminonaphthalene-5-sulphonyl chloride (dansyl chloride) (5 mg/ml) were then added. The tube was stoppered or covered with Parafilm and the mixture was incubated at 37 °C for 2 h or until the yellow colour of the reagent had disappeared. The reaction mixture was then dried down and hydrolysed with 100 μl 6 M HCl for 6–24 h. The hydrolysate was dried down and applied, in 1–2 μl of pyridine/water (1 : 1, v/v) to an origin spot near the corner of a polyamide thin-layer plate (5 cm × 5 cm, double sided). A marker mixture was applied to the origin on the other side of the plate.

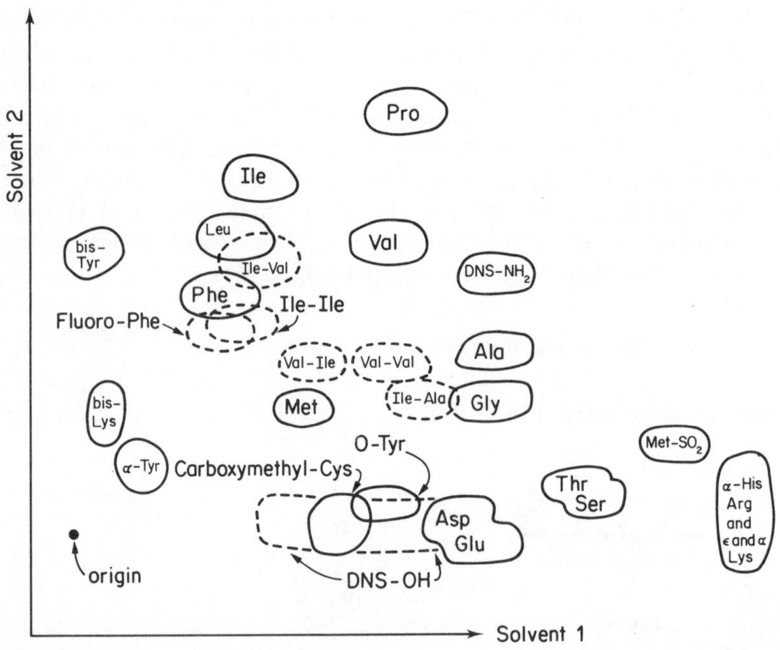

Fig. A1.1. The separation of dansyl amino acids on polyamide thin-layers (Woods and Wang, 1967). Solvent 1: aqueous formic acid (1.5%, v/v). Solvent 2: toluene/acetic acid (9 : 1, v/v).

The chromatogram was developed first by aqueous formic acid (1.5%, v/v) and then, after thorough drying in a stream of cold air, by toluene/acetic acid (9 : 1, v/v). Figure A1.1 shows the position of the normal dansyl amino acids, together with some less common substances, as revealed in ultraviolet light. Figure A1.2 shows the effect of development with other solvents.

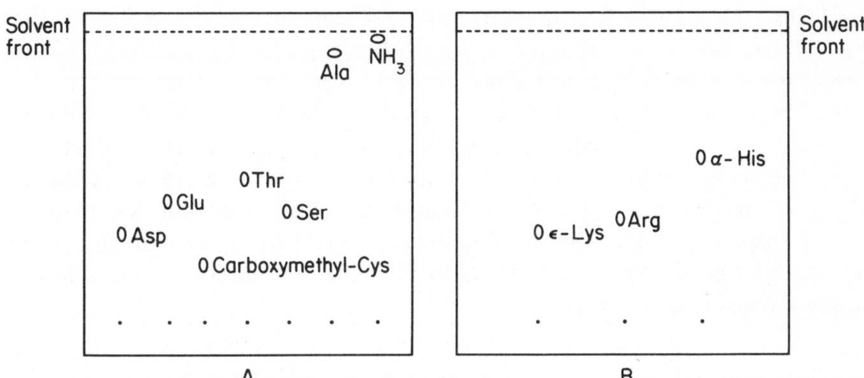

Fig. A1.2. Solvents for the resolution of close pairs in the system of Fig. A1.1. The illustrations are of one-dimensional development of marker substances. (A) Solvent: ethyl acetate/methanol/acetic acid (20 : 1 : 1 by vol.) (Crowshaw *et al.*, 1967). (B) Solvent: 3 vol. pyridine/acetic acid/water (9 : 16 : 1000 by vol.) plus 1 vol. ethanol (Magnusson, S., cited by Jörnvall, 1970). Both these solvents can be used on plates that have already been developed as in Fig. A1.1. The solvent is allowed to ascend in the same direction as that used for Solvent 2 in the original run.

B. *Proteins*. The protein is dissolved (1 mg in 0.2 ml) in 0.5 M $NaHCO_3$/8 M urea and 0.2 ml of a solution of dansyl chloride in acetone (20 mg/ml) added. The mixture was allowed to stand at 37 °C until the yellow colour of the reagent disappeared. The reaction mixture was dialysed (in boiled tubing, see Note 2 to Example (2.6)) and freeze dried in a hydrolysis tube. The hydrolysis was carried out as above, using 100 μl of acid. The quantity of the extract of the dried hydrolysate that was expected to contain 10 nmol of dansyl amino acid was applied for chromatography as above.

Notes

(1) The reagent reacts not only with α-NH_2 groups but also ε-NH_2 groups and the —OH groups of tyrosine. These reactions also give rise to fluorescent products, N^{ε}-dansyl lysine and *O*-dansyl tyrosine. (Amino-terminal lysine and tyrosine produce the bis-dansyl derivatives.) All these products can be distinguished from the α-dansyl compounds, but their production consumes reagent. Proteins, which tend to contain more reactive side-chains in proportion to the α-NH_2 group than do peptides, therefore require a stronger solution of reagent.

(2) The urea in Method B is intended to reduce any steric hindrance of the α-NH_2 group by denaturing the protein. It is also often able to overcome insolubility. Gray (1972) proposes the use of sodium dodecyl sulphonate for such purposes.

(3) The reason for the initial drying down of the peptide in $NaHCO_3$ is to drive off NH_3. Ammonia also consumes reagent and gives rise to a large spot of dansyl-NH_2, which obscures dansyl alanine on the chromatogram. The pH

check that the solution had remained alkaline while drying down ensures that the NH_3 had been removed and that the pH is correct for the next step. Dansyl-NH_2 is removed by dialysis in the procedure as applied to proteins.

(4) The trifluoroacetyl group is sufficiently labile to alkali to be partly deprotected during the dansylation step and give a spurious dansyl product.

(5) Ultraviolet light of both 254 and 356 nm will produce an adequate degree of fluorescence. The plastic foundation of the double-sided plates is fairly transparent to the 356 nm radiation and quite transparent to the visible light emitted by the spots: this leads to difficulties in observing one side free of interference from the other.

Example (A.3). Quantitative determination of free —NH_2 groups

Source. Levy and Chung (1955).

Process involved.

R_1 R_2 R_3 R_4

H_2N—···

\+

F—C_6H_3(O_2N)—NO_2 ⟶ O_2N—C_6H_3(NO_2)—NH—···(R_1 R_2 R_3 R_4)

↓

spectrophotometric estimation ⟵ (selective extraction of N^{α}-derivative) O_2N—C_6H_3(NO_2)—NH—(R_1)—COOH

\+

free amino acids,
side chain
substituted Tyr
and Lys (A.2)

As described here, the Method is used for the determination of side-chain —NH_2 groups. Any N^{α}-dinitrophenyl compounds are removed by ether extraction.

Method

The protein (0.2 μmol) and an equal weight of $NaHCO_3$ were dissolved or suspended in 10 times the weight of water in a test tube. To this was added twice the volume of an ethanolic solution (5%, v/v) of 1-fluoro-2,4-dinitrobenzene and the mixture shaken for 2 h at 20 °C. The mixture was then acidified with a little conc. HCl and shaken with several 25 ml portions of ether. The aqueous suspension of protein was dialysed against several changes of distilled water and freeze dried. The residue was hydrolysed with 6 M HCl (Example (A.1)). The hydrolysate was dried, suspended in water, and extracted with ether. The aqueous layer was adjusted so as to be 1 M in NaOH for spectroscopy. The molar extinction coefficient of N^{ε}-dinitrophenyl lysine is approximately $1{\cdot}7 \times 10^4$ at the peak wavelength (approximately 360 nm).

Notes

(1) This method is cruder and is less popular nowadays than those involving trinitrobenzene sulphonic acid. However, in the experience of several workers in this Laboratory, even the most carefully controlled determinations with the latter reagent are liable to give quite spurious results with the types of materials encountered in this book.

(2) The 1-fluoro-2,4-dinitrobenzene is toxic. The first contact with the skin usually does no visible harm, but the operator may then be sensitized and subsequent exposure can lead to a serious allergic reaction.

Example (A.4). Tryptic digestion (used as a test for racemization)

Source. Rees and Offord (1976b) using standard methods reviewed in Hirs (1967) and by Offord (1969b).

Process involved. Tryptic digestion of a peptide formed by the condensation of two tryptic peptides of hen-egg lysozyme (peptide M34, Table 6.1), followed by examination of the digest on paper.

Method

The peptide (100 nmol) was dissolved in 0.1 ml NH_4HCO_3 (0.5%, w/v) or a suitable semi-aqueous solvent. An aqueous solution of trypsin (2.5 μl, 10 mg/ml) was added and digestion allowed to proceed for 4 h at 37 °C. The digest was dried down and applied for electrophoresis on Whatman 3 MM paper at pH 6.5 (buffer: pyridine/acetic acid/water, 25:1:225 by vol.). The paper was dried and developed by partition chromatography at right angles (solvent: butanol/acetic acid/water/pyridine, 30:6:24:20 by vol.). The paper was dried again and stained with ninhydrin/cadmium acetate. The ninhydrin was 1% (w/v) in acetone and 20 ml of it were mixed with 3 ml of a solution made by diluting 100 ml of cadmium acetate in acetate acid (1%, w/v) with 50 ml water. The spots were allowed to come up at room temperature and their positions compared with those of the fragments used for the condensation and of the condensed product. Typically, there was no visible quantity of undigested product and the only spots were in the positions of the authentic tryptic fragments.

Notes

(1) The trypsin should be treated with a chymotrypsin inhibitor (see Note 1 of Example (3.4)).

(2) Trypsin will often digest an insoluble product, but it cannot be relied upon to do so. The semi-aqueous solvents mentioned in the method are discussed in Example (3.5).

(3) A voltage gradient of 60 V/cm is desirable, but a lower value is acceptable.

(4) The paper can be prepared for the second dimension by turning through a right angle or by sewing the strip as described in the standard treatises.

(5) Most peptides stain red with the Cd–ninhydrin reagent. Those with amino-terminal valine or isoleucine come up more slowly. Those with amino-terminal glycine, Ser-Pro-, and Thr-Pro- are yellow (if they also have lysine the yellow eventually turns to orange). Those with amino-terminal Cys(O_3H), Cys-SO_3H, and asparagine come up yellow and may turn orange. Thr-Thr- gives a brown colour. The Cd reagent is preferred to ninhydrin alone because of the greater range of such colours and because of the apparently improved visual contrast. Even N^{α}-Boc- peptides will give some colour, because of deprotection by the acetic acid as the paper dries.

(6) The paper may be over-stained by one of the following methods (the details of the stains are given in many collections of methods, e.g. Offord (1969b)): the Cl_2 stain for the larger peptides that do not always give a good colour with ninhydrin (especially important in the test for racemization, as we are looking for undigested molecules, which will probably be quite large); the 1-nitroso-2-naphthol test for tyrosine; the 4-dimethylaminobenzaldehyde (Ehrlich) test for tryptophan. The Pauly test for histidine (which is given only poorly by N^{im}-Boc—NH–CH(CF_3)-His) and the Sakaguchi test for arginine are much more effective when carried out on a paper that has not been stained with ninhydrin.

(7) If, instead of the Cd reagent, the paper is stained lightly with much more dilute ninhydrin (e.g. 0.025%, w/v in acetone; heat at 100 °C for 15 min) it is possible to locate the spots and elute them for amino-acid analysis. Damage to the peptide is largely restricted to the amino-terminal residue.

(8) Although it is given here as a test for racemization, the method can of course be used equally as a test for any other reaction, such as N^{ε}-acylation, that will block the action of trypsin.

Example (A.5). Looking for amide loss by electrophoresis

Source. Based on Offord (1966, 1977).

Process involved. The use of a semi-empirical relationship between the mass of a peptide, its net charge, and its mobility.

Method

A peptide of expected sequence Trp-Trp-Cys(SO_3^-)-Asn-Asp-Gly-Arg (mol. wt. = 1016) was isolated from a tryptic digest by ion-exchange chromatography.

On paper electrophoresis two spots were detected, one of mobility −0.25, the other of mobility −0.45 (mobility of Asp = −1.0). Figure A1.3 shows that the latter of these peptides has lost its amide group.

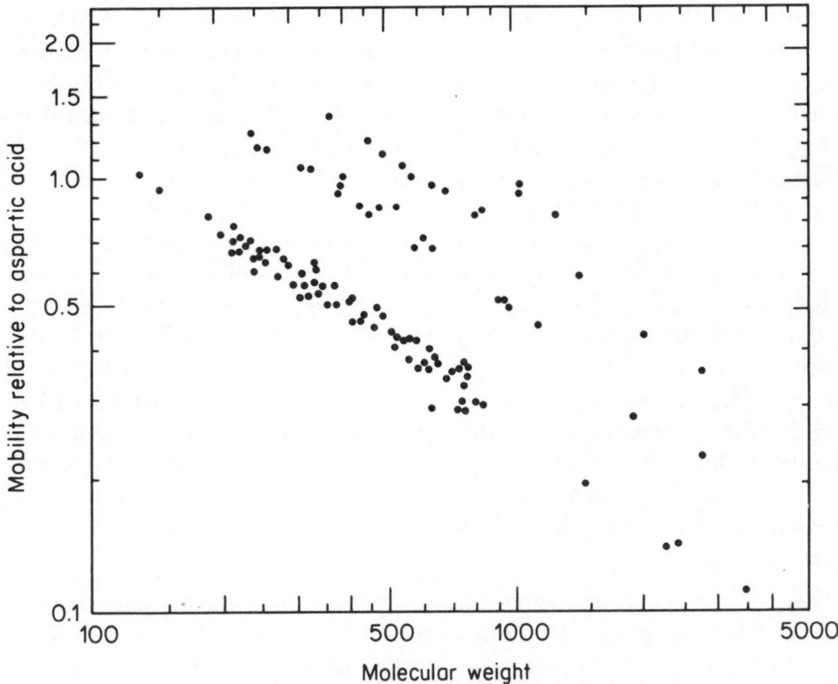

Fig. A1.3. The relationship between electrophoretic mobility of peptides at pH 6.5, their molecular weight and their net charge. The mobilities were determined as described in Note 1 to Example (A.5) and plotted without regard to whether they were anodic or cathodic. Lines of charge of ±1, ±2, ±3, and ±4 are visible on the plot. Peptides that contained histidine and cysteic acid were excluded from the plot. [Reproduced from Offord (1966) by permission of the Publishers of *Nature*.]

Notes

(1) The zero point for the measurement is not the origin line, but the mid-point of a spot of a neutral substance, such as N^{ε}-dinitrophenyl lysine, in order to correct for electroendosmosis. Offord (1966) recommended that the leading edge of the peptide spot should be taken for the mobility measurement as its position is insensitive to overloading.

(2) If the mobility is low, it may be desirable to run the electropherogram for longer than usual. If the aspartic acid standard is run off the paper a secondary standard may be used such as the blue dye xylene cyanol FF ($m = -0.41$). Some workers include small quantities of fluorescent materials in the run to act as markers.

(3) The method may give unduly pessimistic results as some amides can be lost during tryptic digestion.

(4) The method depends on the ionizable groups having charges of +1, 0, or −1 at pH 6.5. The p*K* values of the amino-acid side-chains are such that there are only three common groups of exception to this rule.

The first and most important group are histidine peptides (p*K* of imidazole side-chain 5.5–7, depending on neighbouring groups). Histidine peptides will therefore not necessarily lie on the usual lines and it would only be possible to give upper and lower limits to the number of amide groups.

The α-NH_2 groups usually have p*K*'s above 7.5, and so have an effectively integral charge at pH 6.5. However some, often in peptides that give a yellow or orange colour with Cd–ninhydrin (see Note 5 to Example (A.4)), have somewhat lower p*K* values and as a result the mobility will be slightly less anodic (usually no more than 10%) than predicted. Such peptides (with the exception of those with amino-terminal Cys(O_3H), which show the effect at its most extreme) have been included in Fig. A1.3 and account for some of the scatter. Offord (1977) has reviewed instances in which peptides are a little less cathodic than predicted, suggesting that the α-COOH groups of these peptides have abnormally high p*K* values. This phenomenon is seen with peptides in which Met SO_2 is carboxyl terminal and in which aspartic and glutamic acids are adjacent to a lysine (Ambler and Brown, 1967): it does not seem to give rise to incorrect answers.

Peptides that have undergone transpeptidation between α- and side-chain —COOH groups can give abnormal mobilities. This suggests that, surprisingly, the newly freed α-COOH groups have higher p*K* values than those of the side-chain —COOH groups which are not normally much greater than 5. Transpeptidation is much more clearly revealed on paper electrophoresis at pH 3.5 (buffer: pyridine/acetic acid/water, 1 : 10 : 89 by vol.).

Example (A.6). The use of electrophoretic mobilities to examine the state of protection of a peptide

Source. Rees and Offord (1976a).

Process involved. As for Example (A.4).

Method

Table A.1 shows the electrophoretic observations made on the peptide from the previous Example during the stages of its preparation for use as a carboxyl component in a coupling.

Table A.1

Treatment	Mobility (Asp = −1)	Charge (from Fig. A.2)	Inference
Starting material*	−0.25	−1	Free peptide, one side-chain amide
Method of Example (4.1)	−0.42	−2	Boc- group applied successfully
Method of Example (4.3),	Does not run†	0?	Successful esterification?
(R = p-$CH_3OC_6H_4CH_2$—)	−0.22‡	−1	Loss of amide group after esterification
Method of Example (4.8)§	−0.22	−1	Successful deprotection of α-COOH

* A sample that had not suffered detectable amide loss on preparation: only one spot seen.
† Major spot.
‡ Minor, trailing spot: seen if protected peptide is handled a great deal.
§ Carried out on the sample of peptide to which footnote * applied.

Notes

(1) There are many examples in the publications from this Laboratory of the use of this method in semisynthesis. It is very often used in sequence determination and has been used in conventional peptide synthesis: both applications have been reviewed (Offord, 1977). It is powerful when applied to peptides that will run in the protected state (the peptide illustrated was chosen because it is typical of the middle range of difficulty in this respect). The method is therefore entirely suitable for developing and testing novel methods, but unless improved by the admixture of some solubilizing agent to the buffer (e.g. Example (A.7)) is less effective on larger peptides.

(2) A relationship also exists between composition and R_f on paper partition chromatography (Pardee, 1951) which is on a sounder theoretical footing than the electrophoretic one. Pardee's relationship is sometimes used in sequence determination and has been used at least once in semisynthesis to indicate the state of protection of a peptide (Offord, 1969a).

Example (A.7). The determination of the degree of acylation of insulin by zone electrophoresis in buffered 7 M urea

Source. P. C. Chambers (unpublished work) based on the method of Carpenter and Hayes (1963) for des-amido-insulins.

Process involved. Electrophoresis on a paper or cellulose acetate support in the presence of sufficient urea to prevent aggregation of the protein monomers.

Method

The buffer for the electrophoresis, 0.065 M Na_3PO_4, pH 6.5/7 M urea, was made up by adding 1 vol. 0.52 M Na_3PO_4, pH 6.5 to 7 vol. 8 M urea. The urea was made up as described on p. 217. The sample (approximately 10 nmol) was dissolved in 10 μl of the buffer and applied as a spot, with appropriate markers, to Whatman No. 3 MM paper. The rest of the paper was then wetted in the usual way so that two opposing fronts of buffer met at the origin line. Excess buffer was removed by blotting all parts of the paper except the origin.

Alternatively, of the order of 5–10 nmol was dissolved in approximately 2 μl and applied, 1 μl at a time, to a cellulose acetate strip. If there was enough material, a 10 mg/ml solution was made up and two 1 μl portions applied. The strip had been pre-wetted with the buffer by floating the strip, a portion at a time, on the top of a quantity of buffer kept in a Petri dish. Excess buffer was blotted from the cellulose acetate before application of the sample. Electrophoresis then took place at up to 40 V/cm for the paper and up to 100 V/cm for the cellulose acetate. The strips were removed from the electrophoresis apparatus after a suitable interval, dried in air at 80 °C, and dipped for a few seconds into a solution (0.2%, w/v) of Ponceau S in aqueous acetic acid (3%, v/v). The background colour was reduced by subsequent washing with aqueous acetic acid (5%, v/v). The positions of the spots were compared with those of markers of known degree of acylation.

Notes

(1) Cellulose acetate has the advantage of somewhat greater sensitivity, and (an unusual feature for zonal electrophoresis) ability to permit higher voltage gradients without loss of resolution. Both supports develop a certain amount of heat during the run and this must be carried away by one of the standard devices. We use both the flat-plate type and a tank of the type described by Michl (1951), and widely used in sequence work, in which the coolant is white spirit.

(2) The length of time required for a run can be judged by observing the migration of dye markers. Most bands of acyl insulins lie in the region between the origin and xylene cyanol FF. With the intense voltage gradients possible with a solvent-cooled cellulose acetate strip, the run takes only a very few minutes.

(3) Ponceau S, unlike many dyes, is relatively unaffected by the initial presence of urea.

(4) The buffer must be fresh, as urea breaks down to ammonium cyanate and the pH rises, resulting in a loss of resolution of acyl insulins.

(5) A quantitative version of this method exists (Friesen *et al.*, 1978).

Example (A.8). Isoelectric focussing of derivatives of insulin

Source. Borrás and Offord (1970b), Saunders (1974), and H. T. Storey (private communication), based on Wrigley (1968).

Process involved. Electrophoresis in the presence of a mixture of amphoteric buffers. The result is that a pH gradient is established within the tube, and substances migrate to the region corresponding to their isoelectric point and remain there.

Method

Four solutions were made up. Solution I was *N,N,N′,N′*- tetramethylethylenediamine/water (1 : 9, v/v). Solution II was *N,N′*-methylene bisacrylamide (0.952 g), acrylamide (35.70 g), with water added to 100 ml. Solution III was 8 M urea (p. 217). Solution IV was an aqueous solution (50 mg/ml) of $K_2S_4O_6$. Solutions I and II can be kept for 3 months at 4 °C. Solution III should be freshly deionized and solution IV should be discarded after a week at 4 °C.

The four solutions were de-gassed under a water pump vacuum and solutions I and II mixed in the following proportions, together with 0.60 ml of the appropriate Ampholine buffer (40%, w/v, LKB Ltd.): solution I, 0.08 ml; solution II, 2.52 ml; water, 0.9 ml. The protein sample (50–200 μg) was meanwhile dissolved in 1 ml of solution III. The buffer mixture was then added to the protein solution and thoroughly mixed with 20 μl of solution IV. The mixture was quickly transferred to a plastic tube (4.5 mm internal diameter × 75 mm), stoppered at the bottom with Parafilm. The mixture was added so as almost to fill the tube and 50 μl of water layered on top. The gel set in 1 h at 20 °C.

The Parafilm at the bottom of the tube was pierced in several places and the tube placed in an apparatus for vertical electrophoresis. The upper, anode compartment contained H_3PO_4 (0.2%, v/v) and the lower anode compartment contained ethanolamine (2%, v/v). A voltage was applied to the apparatus, beginning at 50 V and increasing in 50 V steps every 10 min until 300 V was reached. The current should not exceed 2 mA per tube. The run was continued at this voltage for 1–2 more hours.

The power supply was disconnected and the tubes removed from the apparatus. The Parafilm was removed and the gels were expelled, by pressure from a rubber teat fixed over the mouth of the tube, into aqueous trichloroacetic acid (5%, w/v). Bands of coagulated protein soon became visible.

Notes

(1) As little as 0.3 ml of the Ampholine produces acceptable results.

(2) The choice of Ampholine buffer depends on the derivatives that are to be expected. The range pH 5–7 is excellent for unmodified insulins, and mono- and diacyl insulins. The triacyl insulins are best separated from the di-derivatives with the range pH 4–6. Derivatives acylated on the histidine side-chains as well as on the —NH_2 groups do not run. Esters of insulin are best separated on the pH range 3–10, but with the exception of the methyl esters the more hearty esterified products (triester and above) do not run.

(3) The persulphate does not appear to harm the running properties of the insulins.

(4) The tube should be plastic because the gels are not so easy to remove from glass.

(5) The water is layered on the gel mixture to provide a flat top and to prevent air from hindering the polymerization.

(6) The relative intensity of the bands can be estimated by eye. Alternatively, the scanning spectrophotometer can be used for a more quantitative estimate—Fig. A1.4 indicates what can be done.

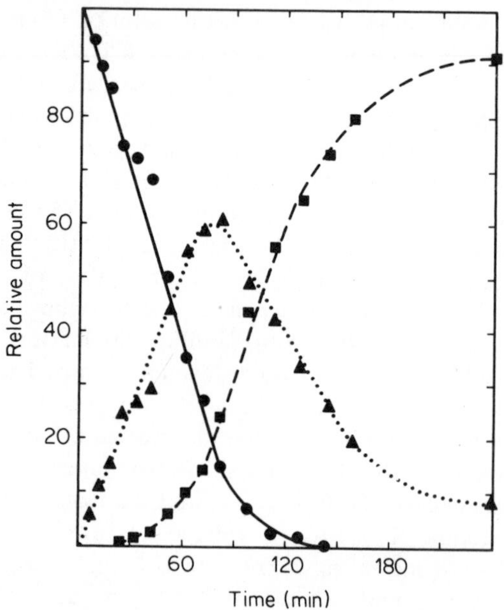

Fig. A1.4. Kinetics of the reaction between insulin (1.5 mM) and phenyl isothiocyanate (12.5 mM) maintained at pH 8.5 in water by the addition of NaOH from a pH-stat. The graph shows the relative intensities of bands obtained on gel isoelectrophoresis of samples taken from the reaction mixture at different times. The symbol (●) indicates unreacted insulin; (▲) indicates monosubstituted insulin; and (■) indicates disubstituted insulin. The intensity of each band is expressed as a percentage of the total intensity on the gel in question. [Reproduced from Borrás (1972) with permission.]

(7) As an alternative to the trichloroacetic acid precipitation the bands can be stained by immersing the gels for 15 min at 60 °C or 60 min at 20 °C in a solution made up as follows (Billcliffe and Arbuthnott, 1974). Coomassie Brilliant Blue R250 (2.5 g) was added to 100 ml of distilled water at 60 °C and stirred vigorously for 10 min. To this was added 100 ml hot (60 °C) 1 M H_2SO_4. After a further 10 min at 60 °C the solution was filtered through Whatman No. 1 paper. The pH of the filtrate was approximately 1.2. Potassium hydroxide (10 M)

was added to pH 5–5.5. Trichloroacetic acid was then added to a final concentration of 12% (w/v). The solution was stored in a brown bottle. The pH should be approximately 1 (readjusted if necessary during storage) and any precipitate filtered off before use. The gels were de-stained in tap water.

(8) Isoelectric focussing, although quite useful, is marred by a tendency to non-reproducible variations in the positions of the bands. Results with it should be treated with caution unless confirmed in some other way.

Example (A.9). Thin-layer chromatography on silica-gel plates

Source. Wünsch (1974).

Process involved. Differential adsorption–desorption of substances of differing degree of hydrophobicity.

Method

The sample (10–40 μg (i.e. about 50–200 nmol) in 2 μl of a suitable solvent) was applied to a plate (silica-gel G on a plastic backing, bought ready-made) and allowed to dry. The plate was then developed, in a tank equipped for ascending chromatography, with chloroform/methanol (9 : 1, v/v). When the front neared the top of the plate the elution was stopped and the plate dried. It was stained either by exposure to I_2 vapour or (after 15 min exposure to the vapour of conc. aqueous HCl, followed by heating to 80 °C for 10 min) with a spray of the Cd–ninhydrin reagent of Example (A.4). Amino acids have $R_f = 0$; acyl amino acids have $R_f = 0.1$–0.2 in the $—COO^-$ form and about 0.5 in the —COOH form; and their active esters have $R_f = 0.6$–0.8.

Notes

(1) The purpose of the exposure to HCl is to remove acid-labile protecting groups from the $—NH_2$ group, in order to obtain a positive result with ninhydrin. Other stains can be used, although the Sakaguchi stain for arginine is remarkably poor, even on an otherwise unstained plate.

(2) If the system is being used to follow the esterification of a compound such as N^{α}-Boc-glutamic acid t-butyl ester, a modified solvent is advisable. The starting compound can, as usual, exhibit a variable R_f, depending on whether or not it is in the carboxyl or carboxylate form (cf. Example (3.7)) but, since it is already highly protected, with a high R_f, the variations lead to its being confused with the esterified product. The addition of 1 vol. *N*-ethylmorpholine to 20 vol. of the above solvent overcomes this problem.

(3) Stewart and Young (1969) provide a great deal of information on the t.l.c. of amino acids and their derivatives. They mention in particular chloroform/methanol/acetic acid (85 : 10 : 5 by vol.) and give R_f values for a number of compounds.

(4) The butanol/acetic acid/water/pyridine solvent mentioned in Example (A.4) is, when used with silica plates, excellent for following —NH_2-protection reactions (S. Hoare, unpublished experiments).

(5) It is often only necessary for the front to move about 4 cm before reactants are sufficiently resolved from products. This provides a very rapid technique for following the extent of a reaction, but it should only be used if appropriate markers are available. The numerical values of R_f are not of themselves at all informative at this stage in the development of the run.

Example (A.10). Chromatography on silica-treated paper

Source. R. R. Moore (1978a).

Process involved. Partition paper chromatography with the system of Waley and Watson (1953), as in Example (A.4). The effect of the silica is to keep the spots sharp and to lessen the trailing of those of high R_f.

Method

Samples (100–200 μg) were dissolved in 10 μl of the developing solvent (below) and applied as spots to the silica-gel-treated paper (Whatman SG81). The paper was developed by ascending chromatography (solvent: n-butanol/glacial acetic acid/water/pyridine (30 : 6 : 12 : 10 by vol.). The front had moved approximately 10–12 cm in 4 h. The paper was dried and stained by an appropriate method.

Notes

(1) The method has been used for derivatives of insulin and peptides derived from trifluoroacetylated cytochrome *c*. Sample R_f's are: insulin, approximately 0.5; mono-Boc-insulin, approximately 0.65; di-Boc-insulin, approximately 0.75; tri-Boc-insulin, approximately 0.85. These values alter somewhat as the solvent ages. The resolution between di- and tri-derivatives of insulin (which are the important derivatives if one is attempting to control a coupling) is not great, but is sufficient. This method is less susceptible to variation than is that given in Example (A.8).

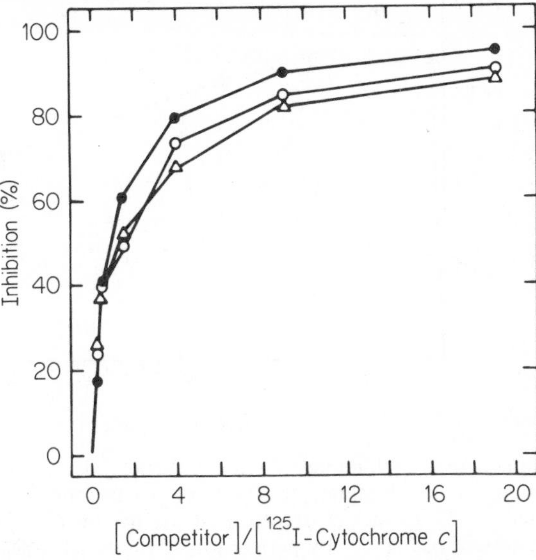

Fig. A1.5. Radioimmunoassay of natural horse cytochrome *c* (●); semisynthetic cytochrome *c* in which residues 1–65 were derived from natural material and residues 66–104 obtained by total synthesis (○); and of cytochrome *c* made by combining natural fragments corresponding to residues 1–65 and 66–104 (Δ). The graph shows the degree of competition of the three materials with ^{125}I-labelled cytochrome *c* for binding to antibody raised against horse cytochrome *c*. [Reproduced from Harbury (1978) by permission. Copyright by Academic Press Inc.]

Fig. A1.6. Crystals of semisynthetic [Thr^{B30}]porcine insulin—i.e. semisynthetic human insulin. The shape and uniformity of the crystals compare very well with those of natural insulin. [Reproduced from Obermeier (1978) by permission. Copyright by Academic Press Inc.]

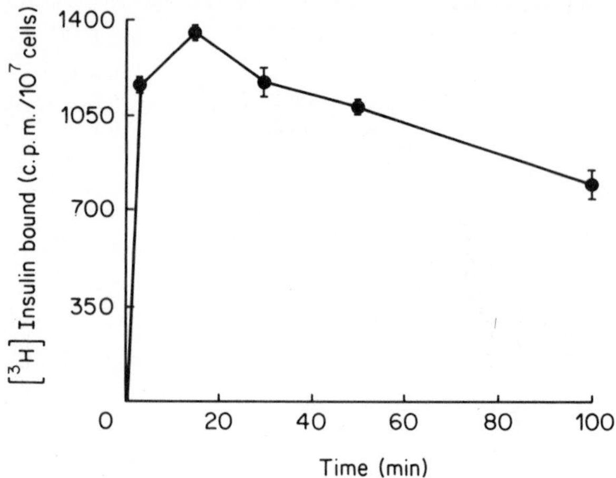

Fig. A1.7. Time-course of binding of semisynthetic [^{3}H]insulin to cultured human lymphocytes. The lymphocytes (between 5×10^6 and 6×10^6 cells/ml) were incubated at 22 °C in the presence of 1.6 nM insulin. [Reproduced from Halban and Offord (1975) by permission of the Biochemical Society.]

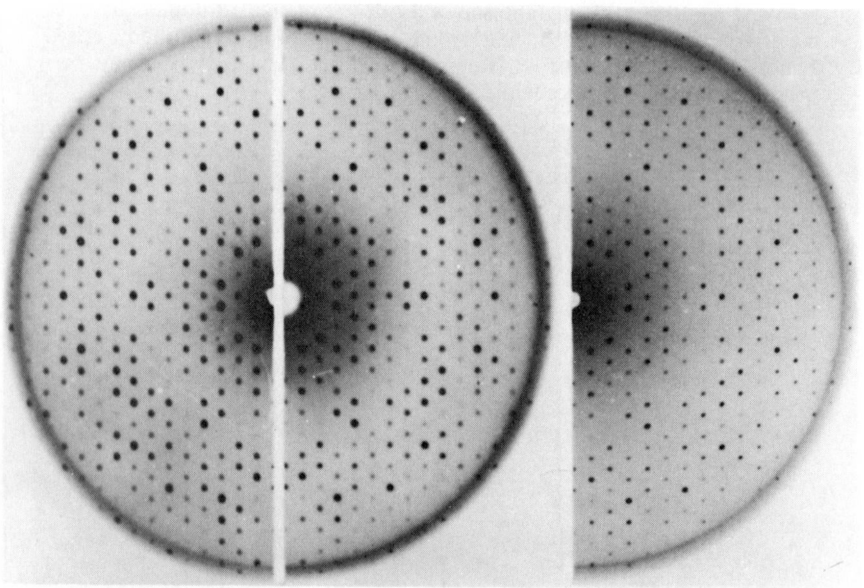

Fig. A1.8. X-ray diffraction patterns of native ribonuclease S′ (left); semi-synthetic ribonuclease S′ of natural sequence (centre); and semisynthetic des-(16-20)-ribonuclease S′ (right). The molecules are clearly very similar in structure. [Reproduced from Chaiken (1978) by permission. Copyright by Academic Press Inc.]

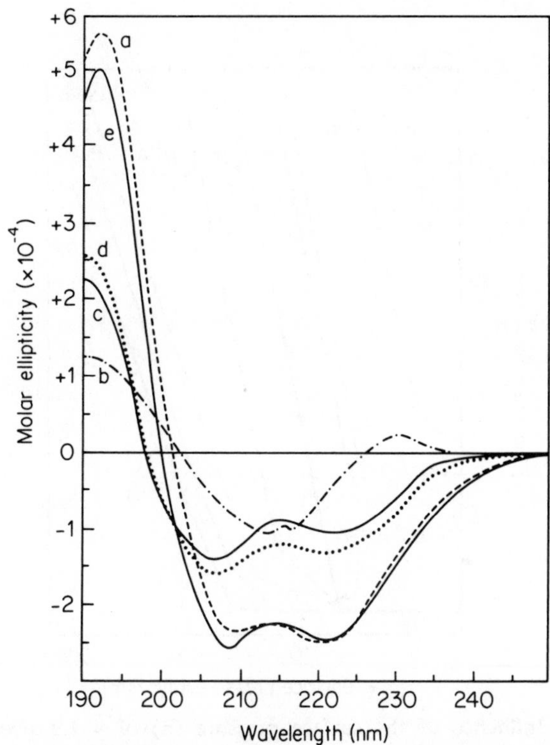

Fig. A1.9. Circular-dichroism spectra (in water at pH 6.0) of myoglobin derivatives: (a) native myoglobin; (b) fragment 1–16; (c) fragment 15–153; (d) fragment 1–16 + fragment 15–153; (e) semisynthetic myoglobin of natural sequence. The semisynthetic and native products are virtually identical. [Reproduced from Wang *et al.* (1978) by permission. Copyright by Academic Press Inc.]

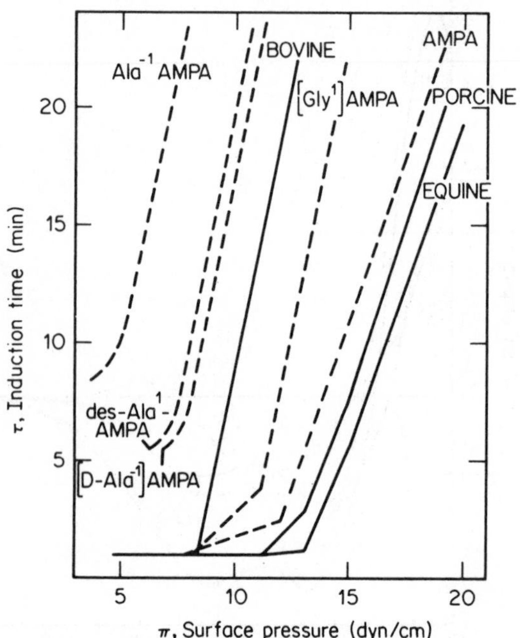

Fig. A1.10. Influence of the surface pressure (π) of a 1,2-didecanoyl-*sn*-glycero-3-phosphorylcholine monolayer film on the induction time (τ) of the enzymic activity of various natural and semisynthetic pancreatic phospholipases A_2. AMPA stands for N^{ε}-acetimidylated porcine pancreatic phospholipase A_2. These measurements give an indication of the enzymes' ability to penetrate water–lipid interfaces—an ability which is essential for them to be able to degrade micellular substrates. [Reproduced from Slotboom *et al.* (1978) by permission. Copyright by Academic Press Inc.]

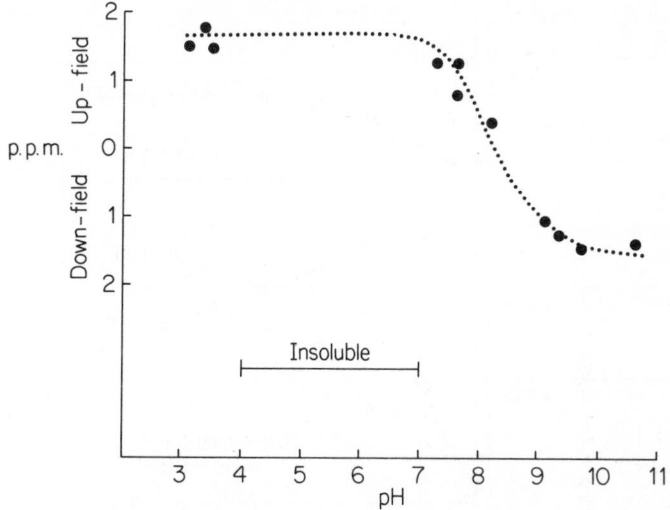

Fig. A1.11. The variation with pH of the peak position of the ^{13}C nuclear magnetic resonance in semisynthetic [Gly^{B1}-2-^{13}C]insulin. The solvent was D_2O/H_2O (1 : 4, v/v). The region of extreme insolubility of insulin is indicated. The chemical shifts are given in parts per million of the frequency of the resonance of carbon atom 2 in free glycine when measured at pH 7 in D_2O/H_2O (1 : 4, v/v). The resonance titrates with an apparent pK of approximately 8.2. [Reproduced from Saunders and Offord (1972), by permission of the North-Holland Publishing Company.]

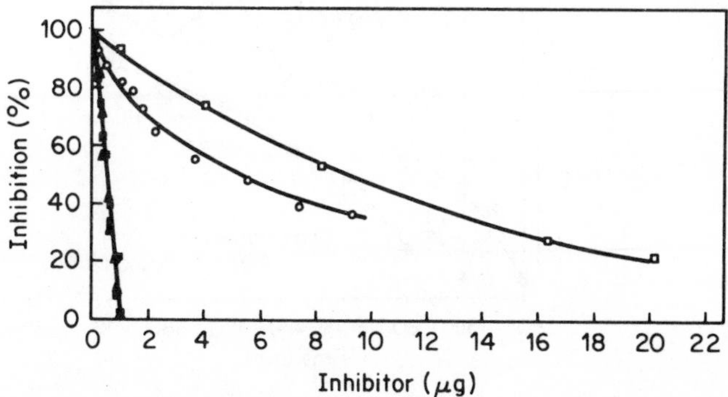

Fig. A1.12. Titrations under identical conditions of bovine trypsin at pH 7.8 in 0.05 M triethanolamine HCl (0.01 M $CaCl_2$) at 25 °C with bovine trypsin-kallikrein inhibitor (Kunitz) and some of its semisynthetic analogues. The symbol (■) indicates the natural inhibitor; (▲) indicates the [Arg^{15}] analogue; (○) indicates the [Phe^{15}] analogue; and (□) the [Trp^{15}] analogue. The [Arg^{15}] analogue has full activity: additional experiments show that the other two analogues were effective against chymotrypsin. [Reproduced from Jering and Tschesche (1978) by permission. Copyright by Academic Press Inc.]

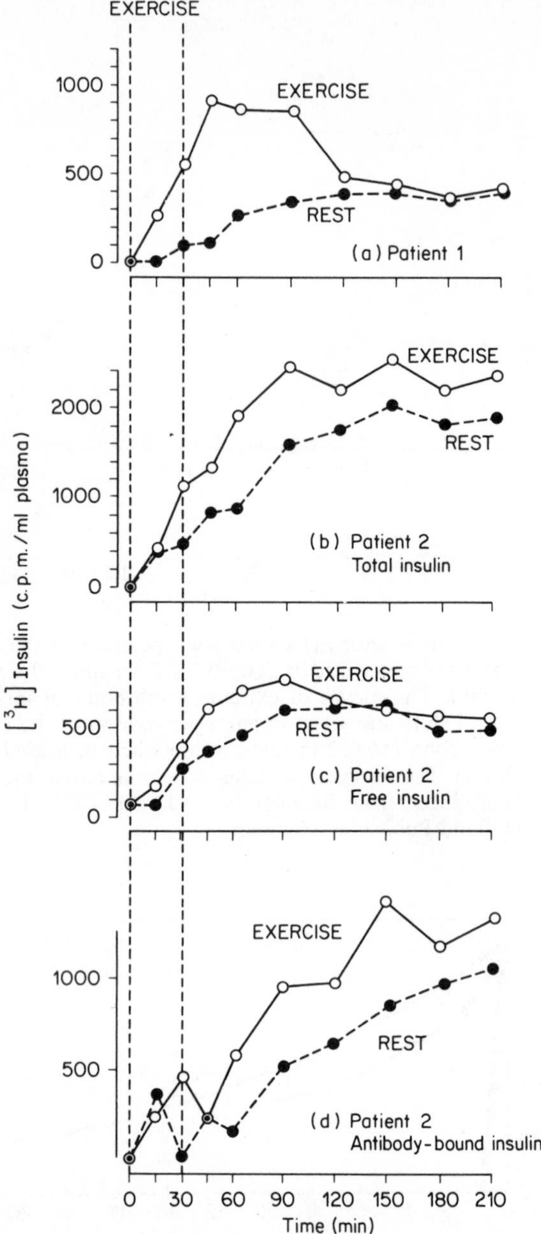

Fig. A1.13. The rate of appearance in the blood of subcutaneously injected semisynthetic [^{3}H]insulin as a function of exercise and the circulating level of antibodies to insulin. (a) Patient 1 (newly diagnosed, no antibodies against commercial insulin). (b) Patient 2 (who had been treated for some time and had acquired antibodies). The insulin levels shown are the sum of those of antibody-bound and free hormone. (c) The level of free insulin in experiment (b). (d) The level of antibody-bound insulin in experiment (b). Antibodies seem in this instance to be associated with an accelerated mobilization of the injected insulin (which partly obscures the effects of exercise). The antibody-bound fraction may be acting as a circulating pool of the hormone. Redrawn from data in Berger *et al.* (1979).

APPENDIX 2

Some common technical problems

This Appendix collects together four miscellaneous methods. The external indication of pH has been mentioned many times in the previous Examples, as has the purification of two important reagents in semisynthetic work, urea and dimethylformamide. I have not included the purification of other reagents, since standard methods will do. The Appendix ends with a note on solubilities, since an otherwise promising semisynthetic route may founder on the lack of solubility of an intermediate at some vital stage.

External indication of pH

The pH in organic solvents was indicated externally as follows. (The method resembles, on a smaller scale, that of Sieber *et al.* (1974).) Narrow-range pH papers (British Drug Houses, Poole, Dorset, U.K.) were moistened with a drop of distilled water, and the drop allowed to soak in completely. The organic solution (0.1–0.5 μl) was discharged from a microcapillary onto a point just inside the wet zone and the colour read after about 5 s. The paper of nominal range 6–8 is most informative when controlling reactions that involve the suppression of the charge on the —NH_2 group, since its effective colour range extends to pH 9. The paper with nominal range 8–10 is only really effective above pH 9 and is thus mainly of use in detecting an over-addition of base.

Used exactly as described, the method gives consistent results and success in the reactions so controlled correlates with one's obtaining the appropriate colour. It does not provide an accurate value of [H^+], but is merely a means of process control.

Purification of urea

Even urea of Analytical Reagent grade has solid impurities, and any solution of urea, particularly if it is stored above neutrality, will generate ammonium cyanate. Accordingly the urea is first dissolved to a concentration of 8 mol/l (480 g in a *final* volume of 1 litre) and filtered through a membrane filtration device (a porosity of 0.2–0.5 μm is required). If this is not possible, one can use a 46 cm × 57 cm sheet of Whatman 3 MM paper, previously macerated in water in a blender, sucked nearly dry on a sheet of No. 1 filter paper in a large Buchner funnel. The filtered solution is stored in a reservoir permanently connected to an ion-exchange column (Amberlite, mixed bed MB3 (British Drug Houses), approximately 3 cm × 60 cm) and drawn off as required at a rate of no more than 10 ml/min. The solution that emerges at first may contain cyanate, as it will have been in the dead volume below the exchanger since the last delivery of urea. Solution with a specific conductivity of less than 1.5 μmhos is satisfactory and can be diluted, if required, to the appropriate concentration. The exchanger is eventually depleted, both by the passage of impure urea solution and the continuous generation of cyanate by the urea solution left within the bed of the column when it is not in use. The conductivity of the eluate then rises.

Purification of dimethylformamide

Dimethylformamide (Laboratory Reagent grade) was stood over K_2CO_3 for 24 h, then filtered and refluxed under reduced pressure over phthalic acid (Analytical Reagent grade) for 1 h. The solvent was then distilled through a fractionating column at 45–55 °C under reduced pressure. For very critical applications such as couplings, the dimethylformamide must be used at once. For less critical applications, such as protection reactions, it was found to be acceptable to store the reagent at +4 °C, with the exclusion of moisture. However, there is a gradual accumulation of formic acid and amine decomposition products which must be removed by evaporation. In spite of its high boiling point, dimethylformamide evaporates very rapidly under the vacuum of the freeze drier. A sample of the solvent should be reduced to two-thirds of its volume to ensure that all the volatile impurities are removed. The remaining liquid is quite cold at the end of this operation, and must therefore be protected from moisture with especial care. It is unwise to treat a sample of dimethylformamide more than once in this way, because of the danger of concentrating a non-volatile impurity.

It is very tedious to redistil the large quantities of dimethylformamide required for the gel-filtration columns described in this book. We have, for non-critical applications, often used Analytical Reagent grade material (after a brief evaporation of volatile impurities) without apparent harm.

The toxicity of dimethylformamide, both by inhalation and skin contact, is frequently underestimated.

A note on solubility

Materials may be thermodynamically or kinetically insoluble in a given solvent. If the problem is thermodynamic, i.e. if the material has a very low solubility even at equilibrium with the solvent, there is nothing to be done except to change the solvent—the biologist cannot raise the temperature significantly, as can most chemists. Solubility in aqueous media can be improved by changes in pH or by the addition of urea, dimethylformamide, or similar compounds. Kinetic insolubility arises when the substance will not dissolve because of unfavourable physical properties—for example, when the solvent and the outer surface of the material combine to form a sticky layer, or when the molecules of the substance have aggregated non-covalently to a very great extent. Such materials can often be brought into solution in solvents that initially appear to be useless by taking them on a temporary excursion to conditions strong enough to bring them into solution: damage can be minimized by working swiftly, and in the cold. If we then return the solution to the original state, substances that were only kinetically insoluble should remain in solution. There are many examples of this approach: temporary changes in pH; the addition and removal of urea; the increase and decrease of ionic strength (e.g. Note 6 to Example (5.3)); the method of Borrás and Offord (1970a, b) for dissolving reluctant proteins in dimethylformamide.

A few additional suggestions on ways to defeat insolubility—some of them rather drastic—are listed on page 399 of Offord and Di Bello (1978).

The solubility problems attendant on the minimal protection schemes of semisynthesis have been discussed at several points in this book. Insolubility is the bugbear of many reactions and of the means of characterizing their products. There are several possible developments in technique that should overcome the problems, but the reader will have noticed that in advance of these developments, recourse is often made to dimethylsulphoxide. This is an excellent solvent but, as has been mentioned, it is probably too reactive to be used without misgivings. Improvements are badly needed.

Bibliography

It may be helpful to have a note on the general literature of the various topics dealt with in this book.

Peptide synthesis. The book by Bodanszky *et al.* (1976) is an excellent summary of the principles of conventional peptide synthesis, and the main ways in which they are put into practice. Schröder and Lübke (1966) produced a more comprehensive survey, but it is now becoming out of date. Law (1974) covers the subject at about the same length as Bodanszky *et al.*, but with much more emphasis on the theoretical aspects of the organic chemistry involved. It occupies a unique position, somewhere between an advanced textbook and a review. Current progress is recorded in the successive volumes: *Amino Acids, Peptides, and Proteins* which appear at yearly intervals as a part of the Specialist Reports of the Chemical Society of the United Kingdom. The conference volumes of the European Peptide Symposia are also invaluable. The full bibliographical references of these and other, similar publications are listed early on in Bodanszky *et al.* (1976). The practical methodology of solid-phase peptide synthesis is comprehensively described by Stewart and Young (1969).

The organic chemistry of peptides, amino acids, and the reagents required for use with them. The collection of recipes and physical data on derivatives by Wünsch (1974) is particularly valuable for its exhaustive listing, with experimental detail, of methods for the preparation of protected, and activated derivatives of amino acids. Greenstein and Winitz (1961) treat a slightly wider field less exhaustively: they cover, in addition to derivatives of amino acids, the chemical reactions of the side-chains, which is useful when contemplating operations of the kind mentioned in Chapter 5. Holton (1971) discusses the preparation of diazoalkanes and gives many detailed examples. Vogel (1967) is a justly famous textbook of general organic-chemical practice.

Protein chemistry. Dayhoff (1972 and later supplements) gives all published amino-acid sequences, both complete and partial. Leggett Bailey (1967) gives, with experimental detail, general techniques for the chemical degradation of proteins. Means and Feeney (1971) and Glazer *et al.* (1976) review the type of chemical modification of proteins that exploits side-chain reactivity. Volumes

11, 25 and 47 of *Methods in Enzymology* (Academic Press) give details of an immense range of techniques for the chemical characterization and sequencing of proteins.

Separation methods. The book by Morris and Morris (1976) gives access to a vast literature on practically every known technique. The volume of *Methods in Enzymology* edited by Jakoby and Wilchek (1974) will be found useful. It is exclusively concerned with affinity methods.

Semisynthesis. Smyth (1975) has written a brief review. Reports of work in progress by most major groups concerned with the technique are collected in Offord and Di Bello (1978).

References

There now follows an alphabetical listing of the references cited throughout the text.

Abita, J. P., Lazdunski, M., Bonsen, P. P. M., Pieterson, W. A. and De Haas, G. H. (1972) *Eur. J. Biochem.*, **30,** 37–47.

Aboderin, A. A., Delpierre, G. R. and Fruton, J. S. (1965) *J. Amer. Chem. Soc.*, **87,** 5469–5472.

Adams, M. J., Blundell, T. L., Dodson, E. J., Dodson, G. G., Vijayan, M., Baker, E. N., Hardin, M. M., Hodgkin, D. C., Rimmer, B. and Sheat, S. (1969) *Nature*, **229,** 491–495.

Akabori, S., Sakakibara, S. and Shunonishi, Y. (1961) *Bull. Chem. Soc. Japan*, **34,** 739.

Akanuma, Y., Kuzuya, T., Hayashi, M., Ide, T. and Kuzuya, N. (1970) *Biochem. Biophys. Res. Commun.*, **38,** 947–953.

Alakhov, Yu. B., Kiryushkin, A. A. and Lipkin, V. M. (1970) *Chem. Commun.*, **1970,** 406.

Ambler, R. P. (1965) *Biochem. J.*, **96,** 32P.

Ambler, R. P. (1967) *Methods Enzymol.*, **11,** 155–166.

Ambler, R. P. and Brown, L. H. (1967). *Biochem. J.* **104,** 784–825.

Anderson, G. W., Zimmerman, J. E. and Callahan, F. M. (1964) *J. Amer. Chem. Soc.*, **86,** 1838–1842.

Anderson, G. W., Zimmerman, J. E. and Callahan, F. M. (1967) *J. Amer. Chem. Soc.*, **89,** 178.

Andersson, B., Nyman, P. O. and Strid, L. (1972) *Biochem. Biophys. Res. Commun.*, **48,** 670–677.

Andrews, P. (1965) *Biochem. J.*, **96,** 595–606.

Atherton, E. and Sheppard, R. C. (1978) *Amino Acids, Peptides, and Proteins*, Vol. 9, Chemical Society Specialist Report, pp. 315–362.

Atherton, E., Clive, D. L. J. and Sheppard, R. C. (1975) *J. Amer. Chem. Soc.*, **97,** 6584.

Axén, R., Porath, J. and Ernback, S. (1967) *Nature*, **214,** 1302–1304.

Backer, T. A. and Offord, R. E. (1968) *Biochem. J.*, **110,** 3P–4P.

Baer, E. and Eckstein, F. (1962) *J. Biol. Chem.*, **237,** 1449–1453.

Banner, D. W., Bloomer, A. C., Petsko, G. A., Phillips, D. C., Pogson, C. I., Wilson, I. A., Corran, P. H., Furth, A. J., Milman, J. D., Offord, R. E., Priddle, J. D. and Waley, S. G. (1975) *Nature*, **255,** 609–614.

Barstow, L. E., Young, R. S., Yakali, E., Sharp, J. J., O'Brien, J. C., Berman, P. W. and Harbury, H. A. (1977) *Proc. Natl Acad. Sci. U.S.A.*, **74,** 4248–4250.

Barton, G. M., Evans, R. S. and Gardner, J. A. F. (1952) *Nature*, **170,** 249–250.

Bates, D. L., Perham, R. N. and Coggins, J. R. (1975) *Anal. Biochem.*, **68,** 175–184.

Bennett, E. L. and Niemann, C. (1950) *J. Amer. Chem. Soc.*, **72,** 1800–1803.

Berger, M., Halban, P. A., Assal, J. P., Offord, R. E., Vranic, M. and Renold, A. E. (1979) *Diabetes*, **28** (Suppl. 1), 53–57.

Bergmann, M. and Zervas, L. (1932) *Chem. Ber.*, 1192–1201.

Berndt, H., Gattner, H.-G. and Zahn, H. (1975) *Hoppe-Seyler's Z. Physiol. Chem.*, **356,** 1469–1472.

Billcliffe, B. and Arbuthnott, J. P. (1974) 'LKB Application Note, May 21, 1974', LKB Ltd., Bromma 1, Sweden. Cited in Arbuthnott, J. P., Billcliffe, B. and Thompson, W. D. (1974) *FEBS Lett.*, **46,** 92–95.

Black, J. A. and Leaf, G. (1965) *Biochem. J.*, **96,** 693–699.

Bláha, K. and Rudinger, J. (1965) *Coll. Czech. Chem.*, **30,** 585–598.

Blake, C. C. F., Koenig, D. F., Mair, G. A., North, A. C. T., Phillips, D. C. and Sarma, V. R. (1965) *Nature*, **206,** 757–761.

Bodanszky, M., Klausner, Y. S. and Ondetti, M. A. (1976) *Peptide Synthesis*, 2nd edn., John Wiley, New York.

de Boer, T. J. and Backer, H. J. (1963) *Org. Syn. Coll.*, **4,** 250–253.

Bolton, A. E. and Hunter, W. M. (1973) *Biochem. J.*, **133,** 529–538.

Boon, P. J., Tesser, G. I. and Nivard, R. J. F. (1978) in Offord and Di Bello (1978) pp. 115–126.

Borrás, F. (1972) *D.Phil. Thesis*, University of Oxford.

Borrás, F. and Offord, R. E. (1970a) *Biochem. J.*, **119,** 24P–25P.

Borrás, F. and Offord, R. E. (1970b) *Nature*, **227,** 716–718.

Borrás, F., Pavani, M., Romo, D. and Malebran, P. (1978) in Offord and Di Bello (1978) pp. 37–43.

Brandenburg, D. (1969) *Hoppe-Seyler's Z. Physiol. Chem.*, **350,** 741–750.

Brandenburg, D. and Ooms, H. A. (1968) *Proc. Int. Symp. Prot. Hormones*, Liège (Ed. Margoulies, M.), Excerpta Med., Amsterdam, pp. 482–484.

Brandenburg, D., Gattner, H. G. and Wollmer, A. (1972) *Hoppe-Seyler's Z. Physiol. Chem.*, **353,** 599–617.

Brandenburg, D., Schermutzki, W. and Zahn, H. (1973) *Hoppe-Seyler's Z. Physiol. Chem.*, **354,** 1521–1524.

Brandenburg, D., Biela, M., Herbertz, L. and Zahn, H. (1975) *Hoppe-Seyler's Z. Physiol. Chem.*, **356,** 961–979.

Braunitzer, G., Beyreuther, K., Fujiki, H. and Schrank, B. (1968) *Hoppe-Seyler's Z. Physiol. Chem.*, **349,** 265.

Bromer, W. W. and Chance, R. E. (1967) *Biochim. Biophys. Acta*, **133,** 219–229.

Bromer, W. W., Knapp-Sheehan, S., Berns, A. W. and Arquilla, E. R. (1967) *Biochemistry*, **6,** 2378–2388.

Bruckner, V., Kótai, A. and Kovács, K. (1959) *Acta Chim. Acad. Sci. Hung.*, **21,** 427–443.

Burton, J. and Lande, S. (1970) *J. Amer. Chem. Soc.*, **92,** 3746–3748.

Busse, W.-D. and Carpenter, F. H. (1976) *Biochemistry*, **15,** 1649–1657.

Busse, W.-D., Hanson, S. R. and Carpenter, F. H. (1974) *J. Amer. Chem. Soc.*, **96,** 5947–5950.

Butler, P. J. G., Harris, J. I., Hartley, B. S. and Leberman, R. (1969) *Biochem. J.*, **112,** 679–689.

Caldwell, J. B., Ledger, R. and Milligan, B. (1966) *Austral. J. Chem.*, **19,** 1297–1298.

Camble, R., Garner, R. and Young, G. T. (1967) *Nature*, **217,** 247–248.

Canfield, R. (1963) *J. Biol. Chem.*, **238,** 2698–2707.

Carpenter, F. H. and Hayes, S. L. (1963) *Biochemistry*, **2,** 1272–1277.

Carpenter, F. H. and Shiigi, S. M. (1974) *Biochemistry*, **13,** 5159–5164.

Carpenter, F. H., Paynovich, R. C. and Yeung, C. W.-T. (1978) in Offord and Di Bello (1978) pp. 219–236.

Carrey, E. A. and Pain, R. H. (1977) *Biochem. Soc. Trans.*, **5,** 689–692.

Chaiken, I. M. (1974) in *Methods Enzymol.* (Eds. Jakoby, W. B. and Wilcheck, M.), Vol. 34, Academic Press, New York, pp. 631–638.
Chaiken, I. M. (1978) in Offord and Di Bello (1978) pp. 349–364.
Chaiken, I. M., Freedman, M. H., Lyerla, J. R., Jr. and Cohen, J. S. (1973) *J. Biol. Chem.*, **248,** 884–891.
Chan, W. W.-C. (1968) *Biochemistry*, **7,** 4247–4254.
Chibnall, A. C., Mangan, J. L. and Rees, M. W. (1958) *Biochem. J.*, **68,** 114–118.
Chillemi, F. and Merrifield, R. B. (1969) *Biochemistry*, **8,** 4344–4346.
Cleland, W. W. (1964) *Biochemistry*, **3,** 480–482.
Cohen, L. A. (1968) *Ann. Rev. Biochem.*, **37,** 695–725.
Collman, J. P. and Buckingham, D. A. (1963) *J. Amer. Chem. Soc.*, **85,** 3039–3040.
Corradin, G. and Chiller, J. M. (1978) in Offord and Di Bello (1978) pp. 91–100.
Corradin, G. and Harbury, H. A. (1970) *Biochim. Biophys. Acta*, **221,** 489–496.
Corradin, G. and Harbury, H. A. (1971) *Proc. Natl Acad. Sci. U.S.A.*, **68,** 3036–3039.
Corradin, G. and Harbury, H. A. (1974a) *Biochem. Biophys. Res. Commun.*, **61,** 1400–1406.
Corradin, G. and Harbury, H. A. (1974b) *Fed. Proc. Fed. Amer. Soc. Exp. Biol.*, **33,** 1302.
Corran, P. H. and Waley, S. G. (1974) *Biochem. J.*, **139,** 1–10.
Cosmatos, A., Okada, Y. and Katsoyannis, P. G. (1976) *Biochemistry*, **15,** 4076–4082.
Cosmatos, A., Cheng, K., Okada, Y. and Katsoyannis, P. G. (1978) *J. Biol. Chem.*, **253,** 6586–6590.
Counsell, R. E., Smith, T. D., Diguilio, W. and Beierwaltes, W. H. (1968) *J. Pharmaceut. Sci.*, **57,** 1958–1961.
Counsell, R. E., Desai, P., Smith, T. D. and Chan, P. S. (1970) *J. Med. Chem.*, **13,** 1040–1042.
Creighton, T. E., Dyckes, D. F. and Sheppard, R. C. (1978) *J. Mol. Biol.*, **119,** 507–518.
Crowshaw, K., Jessup, S. J. and Ramwell, P. W. (1967) *Biochem. J.*, **103,** 79–85.
Curtius, H. (1912) *J. Prakt. Chem.*, **85,** 437–438.
Curtius, T. (1902) *Chem. Ber.*, **35,** 3226–3228.
Curtius, T. (1904) *J. Prakt. Chem.*, **70,** 57–72.
Dayhoff, M. O. (1972) *Atlas of Protein Sequence and Structure*, Vol. 5, 544 pp. (Supplements: No. 1 (1973) 114 pp; No. 2 (1976) 345 pp; No. 3 (1978) 400 pp.) National Biomedical Research Foundation, Georgetown.
Dekker, C. A. and Fruton, J. S. (1948) *J. Biol. Chem.*, **173,** 471–478.
De Meyts, P., van Obberghen, E., Roth, J., Wollmer, A. and Brandenburg, D. (1978) *Nature*, **273,** 504–509.
Di Bello, C. (1975) in *Peptides 1974* (Ed. Wolman, Y.), John Wiley, Chichester, pp. 173–175.
Di Marchi, R. D., Garner, W. H., Wang, C.-C. and Gurd, F. R. N. (1978a) in Offord and Di Bello (1978) pp. 45–57.
Di Marchi, R. D., Garner, W. H., Wang, C.-C., Hanania, C. I. H. and Gurd, F. R. N. (1978b) *Biochemistry*, **17,** 2822–2829.
Dixon, G. H. (1964) *Excerpta Med. Found. Int. Confr.*, **83,** 1207–1215.
Dixon, H. B. F. and Fields, R. (1972) *Methods Enzymol.*, **25,** 409–419.
Dixon, H. B. F. and Perham, R. N. (1968) *Biochem. J.*, **109,** 312–314.
Dixon, H. B. F. and Weitkamp, L. R. (1962) *Biochem. J.* **84,** 462–468.
Dyckes, D. F., Creighton, T. and Sheppard, R. C. (1974) *Nature*, **247,** 202–204.
Dyckes, D. F., Kini, H. and Sheppard, R. C. (1977) *Int. J. Peptide Protein Res.*, **9,** 340–348.
Dyckes, D. F., Creighton, T. E. and Sheppard, R. C. (1978) *Int. J. Peptide Protein Res.*, **11,** 258–268.

Edman, P. (1950) *Acta Chem. Scand.*, **4,** 277–283.
Edman, P. and Begg, G. (1967) *Eur. J. Biochem.*, **1,** 80–91.
Edmundson, A. B. (1965) *Nature*, **205,** 883–887.
Elliott, D. F. and Morris, D. (1960) *Chimia* (*Aarau*), **14,** 373–374.
Eshdat, Y., Dunn, A. and Sharon, N. (1974) *Proc. Natl Acad. Sci. U.S.A.*, **71,** 1658–1662.
Fahrney, D. E. and Gold, A. M. (1963) *J. Amer. Chem. Soc.*, **85,** 997–1000.
Fanger, M. W. and Harbury, H. A. (1965) *Biochemistry*, **4,** 2541–2545.
Fields, R. and Dixon, H. B. F. (1971) *Biochem. J.*, **121,** 587–589.
Finn, F. M. and Hofmann, K. (1965) *J. Amer. Chem. Soc.*, **87,** 645–651.
Fisher, W. R., Taniuchi, H. and Anfinsen, C. B. (1973) *J. Biol. Chem.*, **248,** 3188–3195.
Fölsch, G. (1974) *Chemica Scripta*, **6,** 32–34.
Fölsch, G. (1975) *Chemica Scripta*, **8,** 228.
Fölsch, G. (1978) in Offord and Di Bello (1978) pp. 309–313.
Fones, W. S. and Lee, M. (1954) *J. Biol. Chem.*, **210,** 227–235.
Fontana, A., Veronese, F. M. and Boccu', E. (1971) *Z. Naturforsch.*, **26b,** 314–317.
Fridkin, M. and Goren, H. J. (1971) *Can. J. Chem.*, **49,** 1578–1581.
Friesen, H.-J., Weimann, J., Novak, J. and Brandenburg, D. (1978) in Offord and Di Bello (1978) pp. 161–179.
Furth, A. J., Milman, J. D., Priddle, J. D. and Offord, R. E. (1974) *Biochem. J.*, **139,** 11–25.
Galpin, I. J., Handa, B. K., Kenner, G. W., Moore, S. and Ramage, R. (1976) *J. Chromatog.*, **123,** 237–242.
Galpin, I. J., Jackson, A. G., Kenner, G. W., Noble, P. and Ramage, R. (1978) *J. Chromatog.*, **147,** 424–428.
Galpin, I. J., Chu, K. Y., Hallett, A., Hudson, D., Kenner, G. W., Morgan, B. A., Noble, P., Ramage, R., Seely, J. and Thorpe, W. D. (1979) in *Peptides 1978* (Eds. (Siemion, I. Z. and Kupryszewski, G.), in press, Wydawnictwa Uniwersytetu, Wrocławskiego, Wrocław.
Garner, W. H. and Gurd, F. R. N. (1975) *Biochem. Biophys. Res. Commun.*, **63,** 262–268.
Garner, W. H., Keirn, P., Marshall, R. C., Morrow, J. S., Virscher, R. B. and Gurd, F. R. N. (1973) *Fed. Proc. Fed. Amer. Soc. Exp. Biol.*, **32,** 501.
Gattner, H. G., Schmitt, E. W. and Zahn, H. (1976) in *Peptides 1976* (Ed. Loffet, A.), Editions de l'Université de Bruxelles, pp. 279–284.
Gattner, H. G., Schmitt, E. W. and Naithani, V. K. (1978) in Offord and Di Bello (1978) pp. 181–191.
Gavish, M., Zakut, R., Wilchek, M. and Givol, D. (1978) *Biochemistry*, **17,** 1345–1351.
Geiger, R. (1976) *Chemiker-Zeitung*, **100,** 111–129.
Geiger, R. and Langner, D. (1971) German Patent DBP 2 162; 13.7.1973/16.2.1971: Hoechst AG.
Geiger, R. and Langner D. (1972) German Patent DOS 2 214 405; 4.10.1973/24.3.1972: Hoechst-AG.
Geiger, R. and Langner, D. (1973) *Hoppe-Seyler's Z. Physiol. Chem.*, **354,** 1285–1290.
Geiger, R. and Langner, D. (1975) in *Peptides 1974* (Ed. Wolman, Y.) John Wiley, Chichester, 1975, pp. 173–175.
Geiger, R. and Obermeier, R. (1973) *Biochem. Biophys. Res. Commun.*, **55,** 60–66.
Geiger, R., Schöne, H.-H. and Pfaff, W. (1971) *Hoppe-Seyler's Z. Physiol. Chem.*, **352,** 1487–1490.
Geiger, R., Obermeier, R. and Tesser, G. I. (1975a) *Chem. Ber.*, **108,** 2758–2763.
Geiger, R., Geisen, K. S., Summ, H. D. and Langner, D. (1975b) *Hoppe-Seyler's Z. Physiol. Chem.*, **356,** 1635–1649.

Geiger, R., Geisen, K. S., Regitz, G. and Summ, H.-D. (1976) *Hoppe-Seyler's Z. Physiol. Chem.*, **357**, 1267–1270.

Geiger, R., Teetz, V., König, W. and Obermeier, R. (1978) in Offord and Di Bello (1978) pp. 141–159.

Gerber, J. (1952) *J. Amer. Chem. Soc.*, **74**, 4052–4056.

Glass, J. and Pelzig, M. (1977) *Proc. Natl Acad. Sci. U.S.A.*, **74**, 2739–2741.

Glazer, A. N., Bar-Eli, A. and Katchalski, E. (1962) *J. Biol. Chem.*, **237**, 1832–1838.

Glazer, A. N., De Lange, R. J. and Sigman, D. S. (1976) in *Laboratory Techniques in Biochemistry and Molecular Biology* (Eds. Work, T. S. and Work, E.), Vol. 4, North-Holland, Amsterdam, pp. 1–205.

Gold, A. M. and Fahrney, D. E. (1967) *Biochemistry*, **5**, 2911–2913.

Goldberger, R. F. and Anfinsen, C. B. (1962) *Biochemistry*, **1**, 401–405.

Gray, W. R. (1967) *Methods Enzymol.*, **11**, 139–151.

Gray, W. R. (1972) *Methods Enzymol.*, **25**, 121–138.

Greenstein, J. P. (1957) *Methods Enzymol.* (Eds. Colowick, S. P. and Kaplan, N. O.), Vol. 3, Academic Press, New York, pp. 554–570.

Greenstein, J. P. and Winitz, M. (1961) *The Chemistry of the Amino Acids*, John Wiley, New York.

Gross, E. and Witkop, B. (1961) *J. Amer. Chem. Soc.*, **83**, 1510–1511.

Gutte, B. and Merrifield, R. B. (1971) *J. Biol. Chem.*, **246**, 1922–1941.

Guttman, S. and Boissonnas, R. A. (1958) *Helv. Chim. Acta*, **41**, 1852–1867.

de Haas, G. H., Slotboom, A. J., Bonsen, P. P. M., van Deenen, L. L. M., Maroux, S., Puigserver, A. and Desnuelle, P. (1970a) *Biochim. Biophys. Acta*, **221**, 31–53.

de Haas, G. H., Slotboom, A. J., Bonsen, P. P. M., Nieuwenhuizen, W., van Deenen, L. L. M., Maroux, S., Dlouha, V. and Desnuelle, P. (1970b) *Biochim. Biophys. Acta*, **221**, 54–61.

de Haas, G. H., Bonsen, P. P. M., Pieterson, W. A. and van Deenen, L. L. M. (1971) *Biochim. Biophys. Acta*, **239**, 252–266.

Haber, E. and Krause, R. M. (Eds.) (1977) *Antibodies in Human Diagnosis and Therapy*, Raven Press, New York.

Hagenmeier, H., Ohms, J.-P., Jahns, J. and Anfinsen, C. B. (1978) in Offord and Di Bello (1978) pp. 23–35.

Halban, P. A. and Offord, R. E. (1975) *Biochem. J.*, **151**, 219–225.

Halban, P. A., Karakash, C., Davies, J. G. and Offord, R. E. (1976) *Biochem. J.*, **160**, 409–412.

Halban, P. A., Berger, M., Gjinovci, A., Renold, A., Vranic, M. and Offord, R. E. (1978) in Offord and Di Bello (1978) pp. 237–247.

Hancock, S. and Adams, D. D. (1974) *Chem. in New Zealand*, **38**, 114.

Hanson, H. and Illhardt, R. (1954) *Hoppe-Seyler's Z. Physiol. Chem.*, **298**, 210.

Hantgan, R. R. and Taniuchi, H. (1976) *Fed. Proc. Fed. Amer. Soc. Exp. Biol.*, **35**, 1605.

Hantgan, R. R. and Taniuchi, H. (1977) *J. Biol. Chem.*, **252**, 1367–1374.

Harbury, H. A. (1978) in Offord and Di Bello (1978) pp. 73–89.

Harris, D. E. (1977) *D.Phil. Thesis*, University of Oxford.

Harris, D. E. (1978) in Offord and Di Bello (1978) pp. 127–138.

Harris, D. E. and Offord, R. E. (1977) *Biochem. J.*, **161**, 21–25.

Harris, J. I. (1959) *Nature*, **184**, 167–169.

Hayward, C. H. and Offord, R. E. (1971) in *Peptides 1969* (Ed. Scoffone, E.), North-Holland, Amsterdam, pp. 116–120.

Heinrich, M. R. (1972) in *Molecular Evolution* (Eds. Rohlfing, D. L. and Oparin, A. I.), Plenum Press, New York, pp. 331–339.

Henderson, L. E., Henriksson, D. and Nyman, P. O. (1973) *Biochem. Biophys. Res. Commun.*, **52**, 1388–1394.

Hirs, C. H. W. (Ed.) (1967) *Methods Enzymol.*, **11**, 988 pp.

Hochman, J., Gavish, M., Inbar, D. and Givol, D. (1976) *Biochemistry*, **15**, 2706–2710.
Hofmann, K. and Katsoyannis, P. G. (1963) in *The Proteins* (Ed. Neurath, H.), 2nd edn., Vol. 1, Academic Press, London and New York, pp. 53–188.
Hofmann, K., Smithes, M. J. and Finn, F. M. (1966) *J. Amer. Chem. Soc.*, **88**, 4107–4109.
Hofmann, T. (1964) *Biochemistry*, **3**, 356–364.
Holmgren, A. (1972) *FEBS Lett.*, **24**, 351–354.
Holmgren, A. (1975) *Proc. X FEBS Meeting*, 35–48.
Holton, T. C. (1971) *Ph.D. Thesis*, Ohio State University.
Homandberg, G. A. and Laskowski, M., Jr. (1978) *Fed. Proc. Fed. Amer. Soc. Exp. Biol.*, **37**, 1813.
Homandberg, G. A. and Laskowski, M. Jr., (1979) *Biochemistry*, **18**, 586–592.
Homandberg, G. A., Mattis, J. A. and Laskowski, M., Jr. (1978) *Biochemistry*, **17**, 5220–5227.
Hong, J.-S. and Rabinowitz, J. C. (1970) *J. Biol. Chem.*, **245**, 4988–4994.
Hunter, M. L. and Ludwig, M. J. (1962) *J. Amer. Chem. Soc.*, **84**, 3491–3504.
Hunter, W. M. (1973) in *Immunochemistry* (Ed. Weir, D. M.), Blackwells, Oxford, pp. 17.1–17.36.
Inouye, K. and Watanabe, K. (1977) *J. Chem. Soc. Perkin Trans.*, **1977**, 1905–1911.
Insulin Research Groups, Pekin and Shanghai (1974) *Scientica Sinica*, **17**, 779–792.
Insulin Research Group, Shanghai (1966) *Kexue Tongbao*, **17**, 241–245.
Insulin Research Group, Shanghai (1973) *Scientica Sinica*, **16**, 61–76.
Iselin, B. (1962) *Helv. Chim. Acta*, **45**, 1510–1515.
Isowa, Y., Ohmori, M., Ichikawa, T., Kurita, H., Sato, M. and Mori, K. (1977) *Bull. Chem. Soc. Japan*, **50**, 2762–2765.
Izumiya, N., Noda, K. and Anfinsen, C. B. (1971) *Arch. Biochem. Biophys.*, **144**, 237–244.
Jakoby, W. B. and Wilchek, M. (Eds.) (1974) *Methods Enzymol.*, **34**, 810 pp.
Jering, H. and Tschesche, H. (1974) *Angew. Chem. Internat. Ed.*, **13**, 660–661.
Jering, H. and Tschesche, H. (1976a) *Eur. J. Biochem.*, **61**, 443–452.
Jering, H. and Tschesche, H. (1976b) *Eur. J. Biochem.*, **61**, 453–463.
Jering, H. and Tschesche, H. (1978) in Offord and Di Bello (1978) pp. 283–298.
Jollès, J., Jauregúi-Adell, J., Bernier, I. and Jollès, P. (1963) *Biochim. Biophys. Acta*, **78**, 668–689.
Jörnvall, H. (1970) *Eur. J. Biochem.*, **14**, 521–534.
Kappeler, H. (1963) in *Peptides 1962* (Ed. G. T. Young), Pergamon Press, Oxford.
Kassell, B. and Laskowski, M. (1965) *Biochem. Biophys. Res. Commun.*, **20**, 463–468.
Kato, T. and Tominaga, N. (1970) *FEBS Lett.*, **10**, 313–316.
Kidd, D. A. and King, F. E. (1948) *Nature*, **162**, 776.
Kirby, T. W. and Warme, P. K. (1978) *Anal. Biochem.*, **85**, 367–376.
Kloss, G. and Schroder, E. (1964) *Hoppe-Seyler's Z. Physiol. Chem.*, **336**, 248–256.
Koide, T., Tsunasawa, S. and Ikenaka, T. (1972) *J. Biochem. (Tokyo)*, **71**, 165–167
König, W. and Geiger, R. (1970) *Chem. Ber.*, **103**, 788–798.
König, W. and Geiger, R. (1972) in *Chemistry and Biology of Peptides* (Ed. Meienhofer, J.), Ann Arbor Science Publishers Inc., Michigan, pp. 343–350.
Kovács, J., Medzihradszky, K. and Bruckner, V. (1955) *Acta Chim. Acad. Sci. Hung.*, **6**, 183–189.
Kowalski, D. (1978) in Offord and Di Bello (1978) pp. 263–282.
Kowalski, D. and Laskowski, M., Jr. (1976a) *Biochemistry*, **15**, 1300–1309.
Kowalski, D. and Laskowski, M., Jr. (1976b) *Biochemistry*, **15**, 1309–1315.
Krail, G., Brandenburg, D. and Zahn, H. (1971) *Hoppe-Seyler's Z. Physiol. Chem.*, **352**, 1595–1598.
Krail, G., Brandenburg, D. and Zahn, H. (1973) *Hoppe-Seyler's Z. Physiol. Chem.*, **354**, 1497–1498.

Krail, G., Brandenburg, D. and Zahn, H. (1975a) *Hoppe-Seyler's Z. Physiol. Chem.*, **356,** 981–996.
Krail, G., Brandenburg, D. and Zahn, H. (1975b) *Makromol. Chem., Suppl.*, **1,** 7–22.
Kusama, K. (1957) *J. Biochem.* (*Tokyo*), **44,** 375–381.
Lande, S. (1971) *J. Org. Chem.*, **36,** 1267–1270.
Lande, S. and Burton, J. (1971) in *Peptides 1969* (Ed. Scoffone, E.), North-Holland Publishing Company, Amsterdam, pp. 109–112.
Laskowski, M., Jr. (1978) in Offord and Di Bello (1978) pp. 255–262.
Law, H. D. (1974) *The Organic Chemistry of Peptides*, Wiley–Interscience, London.
Leary, T. R. and Laskowski, M., Jr. (1973) *Fed. Proc. Fed. Amer. Soc. Exp. Biol.*, **32,** 465.
Ledden, D. J., Nix, P. T. and Warme, P. K. (1977) *Fed. Proc.*, **36,** 865.
Leggett Bailey, J. (1967) *Techniques in Protein Chemistry*, 2nd edn., Elsevier, Amsterdam.
Levy, A. L. and Chung, D. (1955) *J. Amer. Chem. Soc.*, **77,** 2899–2900.
Levy, D. (1973) *Biochim. Biophys. Acta*, **328,** 107–113.
Levy, D. and Carpenter, F. H. (1967) *Biochemistry*, **6,** 3559–3568.
Levy, D. and Paselk, R. A. (1973) *Biochem. Biophys. Acta*, **310,** 398–405.
Li, C. H. and Bewley, J. A. (1976) *Proc. Natl Acad. Sci. U.S.A.*, **73,** 1476–1479.
Li, C. H., Blaker, J. and Hayashida, T. (1978) *Biochem. Biophys. Res. Commun.*, **82,** 217–222.
Lin, M. C., Gutte, B., Moore, S. and Merrifield, R. B. (1970) *J. Biol. Chem.*, **245,** 5169–5170.
Lindsay, D. G. (1972) *FEBS Lett.*, **21,** 105–108.
Lode, E. T. (1973) *Fed. Proc. Fed. Amer. Soc. Exp. Biol.*, **32,** 542.
Lode, E. T., Murray, C. L., Sweeney, W. V. and Rabinowitz, J. C. (1974a) *Proc. Natl Acad. Sci. U.S.A.*, **71,** 1361.
Lode, E. T., Murray, C. L. and Rabinowitz, J. C. (1974b) *Biochem. Biophys. Res. Commun.*, **61,** 163–169.
Lode, E. T., Murray, C. L. and Rabinowitz, J. C. (1976a) *J. Biol. Chem.*, **251,** 1675–1682.
Lode, E. T., Murray, C. L. and Rabinowitz, J. C. (1976b) *J. Biol. Chem.*, **251,** 1683–1687.
Ludwig, M. L. and Byrne, R. (1962) *J. Amer. Chem. Soc.*, **84,** 4160–4162.
Ludwig, M. L. and Hunter, M. J. (1967) *Methods Enzymol.*, **11,** 595–604.
Mabery, C. F. and Robinson, F. C. (1882) *Amer. Chem. J.*, **4,** 101–103.
Mahony, P. C. and Offord, R. E. (1977) *Biochem. J.*, **167,** 287.
Marbach, P. and Rudinger, J. (1974) *Helv. Chim. Acta*, **57,** 403–414.
Margoliash, E. and Schejter, A. (1966) *Adv. Protein Chem.*, **21,** 113–286.
Margoliash, E., Smith, E. L., Kreil, G. and Tuppy, H. (1961) *Nature*, **192,** 1125–1127.
Means, G. E. and Feeney, R. E. (1971) *Chemical Modification of Proteins*, Holden-Day, San Francisco.
Michl, H. (1951) *Monatshefte Chem.*, **82,** 489–493.
Mikes, O., Holeysovsky, V., Tomasek, V. and Sorm, F. (1966) *Biochem. Biophys. Res. Commun.*, **24,** 346–352.
Miller, J. B. (1959) *J. Org. Chem.*, **24,** 560–561.
Milman, J. D. (1974) *D.Phil. Thesis*, University of Oxford.
Milne, H. B. and Carpenter, F. H. (1968) *J. Org. Chem.*, **33,** 4476–4479.
Milne, H. B. and Most, C. F. (1968) *J. Org. Chem.*, **33,** 169.
Milne, H. B., Halver, J. E., SoHo, D. and Mason, M. S. (1957) *J. Amer. Chem. Soc.*, **79,** 637–539.
Moore, R. R. (1978a) *D.Phil. Thesis*, University of Oxford.
Moore, R. R. (1978b) in Offord and Di Bello (1978) pp. 368–371.
Moore, R. R. and Offord, R. E. (1979), in press.

Morihara, K. and Oka, T. (1977) *Biochem. J.*, **163,** 531–542.

Moroder, L., Marchiori, F., Rocchi, R., Fontana, A. and Scoffone, E. (1969) *J. Amer. Chem. Soc.*, **91,** 3921–3926.

Moroder, L., Hallett, A., Wunsch, E., Keller, O. and Wersin, G. (1976) *Hoppe-Seyler's Z. Physiol. Chem.*, **357,** 1651–1653.

Morris, C. J. O. R. and Morris, P. (1976) *Separation Methods in Biochemistry*, Pitman, London.

Mosbach, K. (Ed.) (1976) *Methods Enzymol.*, **44,** 999 pp.

Neet, K. E. and Koshland, D. E. (1966) *Proc. Natl. Acad. Sci. U.S.A.*, **56,** 1606–1611.

Nefkens, G. H. L., Tesser, G. I. and Nivard, R. F. (1960) *Rec. Trav. Chim. Pays-Bas*, **79,** 688–698.

Nureddin, A. and Inagami, T. (1969) *Biochem. Biophys. Res. Commun.*, **36,** 999–1005.

Nureddin, A. and Inagami, T. (1975) *Biochem. J.*, **147,** 71–81.

Obermeier, R. (1978) in Offord and Di Bello (1978) pp. 201–211.

Obermeier, R. and Geiger, R. (1976) *Hoppe-Seyler's Z. Physiol. Chem.*, **357,** 759–767.

Offord, R. E. (1966) *Nature*, **211,** 591–593.

Offord, R. E. (1969a) *Nature*, **221,** 37-40.

Offord, R. E. (1969b) in *Data for Biochemical Research* (Eds. Dawson, R. M. C., Elliot, D. C., Elliott, W. H. and Jones, K. M.), 2nd edn., Clarendon Press, Oxford, pp. 525–535.

Offord, R. E. (1972) *Biochem. J.*, **129,** 499–501.

Offord, R. E. (1973) in *Peptides 1972* (Eds. Hanson, H. and Jakubke, H.-D.), North-Holland, Amsterdam, pp. 52–56.

Offord, R. E. (1977) *Methods Enzymol.*, **47,** 51–69.

Offord, R. E. and Di Bello, C. (Eds.) (1978) *Semisynthetic Peptides and Proteins*, Papers presented at the international meeting on protein semisynthesis, September 1977. Academic Press, London and New York.

Offord, R. E., Storey, H. T., Hayward, C. F., Johnson, W. H., Pheasey, M. H., Rees, A. R. and Wightman, D. A. (1976) *Biochem. J.*, **159,** 480–486.

Ontjes, D. and Anfinsen, C. B. (1969) *J. Biol. Chem.*, **244,** 6316–6322.

Overberger, C. G., Weinshenker, N. and Anselme, J.-P. (1965) *J. Amer. Chem. Soc.*, **87,** 4119–4124.

Packer, E. L., Sternlicht, H., Lode, E. T. and Rabinowitz, J. C. (1975) *J. Biol. Chem.*, **250,** 2062–2072.

Pandin, M., Padlan, E. A., Di Bello, C. and Chaiken, I. M. (1976) in *Peptides 1976* (Ed. Loffet, A.), Editions de l'Université de Bruxelles, pp. 319–322.

Pardee, A. B. (1951) *J. Biol. Chem.*, **190,** 757–762.

Parikh, I., Corley, L. and Anfinsen, C. B. (1971) *J. Biol. Chem.*, **246,** 7392–7397.

Paselk, R. A. and Levy, D (1974) *Biochem. Biophys. Acta*, **359,** 215–221.

Pechère, J.-F. and Bertrand, R. (1977) *Methods Enzymol.*, **47E,** 149–155.

Photaki, I. and Bradakos, V. (1965) *J. Amer. Chem. Soc.*, **87,** 3489–3492.

Plapp, B. V. and Kim, J. C. (1974) *Anal. Biochem.*, **62,** 291–294.

Polgar, L. and Bender, M. L. (1966) *J. Amer. Chem. Soc.*, **88,** 3153–3154.

Previero, A., Coletti-Previero, M. A. and Cavadore, J.-C. (1967) *Biochim. Biophys. Acta*, **147,** 453–461.

Radenhousen, R. (1895) *J. Prakt. Chem.*, **52,** 433–454.

Radin, N. S. and Metzler, D. E. (1955) *Biochem. Prep.*, **4,** 60–62.

Rall, S. C., Bollinger, R. E. and Cole, R. D. (1969) *Biochemistry*, **8,** 2486–2496.

Rees, A. R. (1974) *M.Sc. Thesis*, University of Oxford.

Rees, A. R. (1976) *D.Phil. Thesis*, University of Oxford.

Rees, A. R. and Offord, R. E. (1972) *Biochem. J.*, **130,** 965–968.

Rees, A. R. and Offord, R. E. (1976a) *Biochem. J.*, **159,** 467–479.

Rees, A. R. and Offord, R. E. (1976b) *Biochem. J.*, **159,** 487–493.

Reynolds, J. H. (1968) *Biochemistry*, **7,** 3131–3135.

Richards, F. M. (1958) *Proc. Natl. Acad. Sci. U.S.A.*, **44**, 162–166.
Richards, F. M. and Vithayathil, P. J. (1959) *J. Biol. Chem.*, **234**, 1459–1465.
Riley, M. and Perham, R. N. (1970) *Biochem. J.*, **118**, 733–739.
Robinson, N. C., Neurath, H. and Walsh, K. A. (1973) *Biochemistry*, **12**, 420–426.
Rose, K. and Priddle, J. D. (1978) in Offord and Di Bello (1978) pp. 387–395.
Ruttenberg, M. A. (1972) *Science*, **177**, 623–636.
Ryle, A. P., Sanger, F., Smith, L. F. and Kitai, R. (1955) *Biochem. J.*, **60**, 541–556.
Sano, S. and Kurihara, M. (1969) *Hoppe-Seyler's Z. Physiol. Chem.*, **350**, 1183.
Sarin, P. S. and Fasman, G. D. (1964) *Biochim. Biophys. Acta*, **82**, 175–178.
Sarkar, B., Dixon, H. B. F. and Webster, D. (1978) *Biochem. J.*, **173**, 895–897.
Saunders, D. J. (1974) *D.Phil. Thesis*, University of Oxford.
Saunders, D. J. (1978) in Offord and Di Bello (1978) pp. 213–218.
Saunders, D. J. and Offord, R. E. (1972) *FEBS Lett.*, **26**, 286–288.
Saunders, D. J. and Offord, R. E. (1977a) *Biochem. J.*, **165**, 479–486.
Saunders, D. J. and Offord, R. E. (1977b) *Hoppe-Seyler's Z. Physiol. Chem.*, **358**, 1469–1474.
Schlichtkrull, J. (1956) *Acta Chem. Scand.*, **10**, 1455–1458.
Schlichtkrull, J. (1958) *Insulin Crystals*, Munksgaard, Copenhagen.
Schnabel, E. (1967) *Liebigs Ann. Chem.*, **702**, 188–196.
Schnabel, E., Stoltefuss, J., Offe, H. A. and Klanke, E. (1971) *Justus Liebigs Ann. Chem.*, **743**, 57–68.
Schröder, E. and Lübke, K. (1966) *The Peptides*, Vols. I and II, Academic Press, New York.
Schwert, G. W., Neurath, H., Kaufman, S. and Snoke, J. E. (1948) *J. Biol. Chem.*, **172**, 221–239.
Schwyzer, R., Sieber, P. and Kappeler, H. (1959) *Helv. Chim. Acta*, **42**, 2622–2624.
Scoffone, E., Rocchi, R., Marchiori, F., Moroder, L., Marzotto, A. and Tamburro, A. M. (1967) *J. Amer. Chem. Soc.*, **89**, 5450–5455.
Sealock, R. W. and Laskowski, M., Jr. (1969) *Biochemistry*, **8**, 3703–3710.
Sealock, R. W. and Laskowski, M., Jr. (1973) *Biochemistry*, **12**, 3139–3146.
Sheehan, J. C. and Hess, G. P. (1955) *J. Amer. Chem. Soc.*, **77**, 1067–1068.
Shimonishi, Y. (1970) *Bull. Chem. Soc. Japan*, **43**, 3251–3255.
Sieber, P. and Riniker, B. (1973) in *Peptides 1971* (Ed. Nesvadba, H.), North-Holland, Amsterdam, pp. 49–53.
Sieber, P., Kamber, B., Hartmann, A., Jöhl, A., Riniker, B. and Rittel, W. (1974) *Helv. Chim. Acta*, **57**, 2617–2621.
Sieber, P., Kamber, B., Hartmann, A., Jöhl, A., Riniker, B. and Rittel, W. (1977) *Helv. Chim. Acta*, **60**, 27–37.
Slaby, I. and Holmgren, A. (1975) *J. Biol. Chem.*, **250**, 1340–1347.
Slotboom, A. J. and de Haas, G. H. (1975) *Biochemistry*, **14**, 5394–5399.
Slotboom, A. J., van Dam-Mieras, M. C. F. and de Haas, G. H. (1977) *J. Biol. Chem.*, **252**, 2948–2951.
Slotboom, A. J., Jansen, E. H. J. M., Pattus, F. and de Haas, G. H. (1978) in Offord and Di Bello (1978) pp. 315–348.
Smyth, D. G. (1975) *Nature*, **256**, 699–700.
Smyth, D. G., Stein, W. H. and Moore, S. (1963) *J. Biol. Chem.*, **238**, 227–234.
Spero, L., Griffin, B. Y., Middlebrook, J. L. and Metzger, J. F. (1976) *J. Biol. Chem.*, **251**, 5580–5588.
Stahman, M. A. and Becker, R. R. (1952) *J. Amer. Chem. Soc.*, **74**, 2695–2696.
Stelakatos, G. C. and Argyroupoulos, N. (1970) *J. Chem. Soc.*, **1970c**, 964–967.
Stelakatos, G. C., Paganou, A. and Zervas, L. (1966) *J. Chem. Soc.*, **1966c**, 1191–1199.
Stewart, J. M. and Young, J. D. (1969) *Solid-phase Peptide Synthesis*, W. H. Freeman and Co., San Francisco, pp. 40–41.
Swan, J. M. (1957) *Nature*, **180**, 643–645.

Takahashi, K. (1965) *J. Biol. Chem.*, **240,** PC4117–PC4119.
Tanaka, M., Nakashuma, T., Benson, A., Mower, H. and Yasunobu, K. T. (1966) *Biochemistry*, **5,** 1666–1681.
Taniuchi, H. and Anfinsen, C. B. (1971) *J. Biol. Chem.*, **246,** 2291–2301.
Taniuchi, H., Anfinsen, C. B. and Sodja, A. (1967) *Proc. Natl Acad. Sci. U.S.A.*, **58,** 1235–1242.
Taniuchi, H., Davies, D. R. and Anfinsen, C. B. (1972) *J. Biol. Chem.*, **247,** 3362–3364.
Teetz, V., Eckert, H. G. and Geiger, R. (1976) in *Peptides 1976* (Ed. Laffet, A.), Editions de l'Université de Bruxelles, pp. 263–267.
Tesser, G. I. (1975) in *Peptides 1974* (Ed. Wolman, Y.), John Wiley, New York, pp. 53–56.
Van Nispen, J. W., Smeets, P. J. H., Poll, E. H. A. and Tesser, G. I. (1977) *Int. J. Peptide Protein Res.*, **9,** 203–212.
Veber, D. F., Milkowski, J. D., Varga, S. L., Denkewalter, R. C. and Hirschmann, R. (1972) *J. Amer. Chem. Soc.*, **94,** 5456–5461.
Vine, W. H., Breuckner, D. A., Needleman, P. and Marshall, G. R. (1973) *Biochemistry*, **12,** 1630–1637.
Visser, S., Raap, J., Kerling, K. E. T. and Havinga, E. (1970) *Rec. Trav. Chim. Pays-Bas*, **89,** 863–870.
Vogel, A. I. (1967) *A Textbook of Practical Organic Chemistry*, 3rd edn., Longmans, London.
Waldschmidt-Lietz, E. and Kühn, K. (1951) *Chem. Ber.*, **84,** 381–384.
Waley, S. G. and Watson, J. (1953) *Biochem. J.*, **55,** 328–337.
Wallace, C. J. A. (1976) *D.Phil. Thesis*, University of Oxford.
Wallace, C. J. A. (1978) in Offord and Di Bello (1978) pp. 101–114.
Wallace, C. J. A. and Offord, R. E. (1979) *Biochem. J.*, **179,** 169–182.
Walsh, K. and Neurath, H. (1964) *Proc. Natl Acad. Sci. U.S.A.*, **52,** 884–889.
Wang, C.-C., DiMarchi, R. D. and Gurd, F. R. N. (1978) in Offord and Di Bello (1978) pp. 59–69.
Webster, D. (1974) *D.Phil. Thesis*, University of Oxford.
Webster, D. J. and Offord, R. E. (1972) *Biochem. J.*, **130,** 315–317.
Webster, D. and Offord, R. E. (1977) *Biochem. J.*, **167,** 285–287.
Weinert, M., Brandenburg, D. and Zahn, H. (1969) *Hoppe-Seyler's Z. Physiol. Chem.*, **350,** 1556–1562.
Weinert, M., Kircher, K., Brandenburg, D. and Zahn, H. (1971) *Z. Physiol. Chem.*, **352,** 719–724.
Weitzel, G., Bauer, F.-U. and Eisele, K. (1976) *Hoppe-Seyler's Z. Physiol. Chem.*, **357,** 187–200.
Weitzel, G., Bauer, F.-U. and Rehe, A. (1978) in Offord and Di Bello (1978) pp. 193–200.
Weizmann, M. and Patai, S. (1946) *J. Amer. Chem. Soc.*, **68,** 150–151.
Weygand, F. and Csendes, E. (1952) *Angew. Chem.*, **64,** 136.
Weygand, F., Geiger, R. and Glöckler, U. (1956) *Chem. Ber.*, **89,** 1543–1549.
Weygand, F., Steglich, W. and Pietta, P. (1967) *Chem. Ber.*, **100,** 3841–3849.
Wieland, T. and Sehring, R. (1958) *Annalen.*, **569,** 122–124.
Wilgus, H. and Stellwagen, E. (1974) *Proc. Natl Acad. Sci. U.S.A.*, **71,** 2892–2894. *Fed. Proc. Fed. Amer. Soc. Exp. Biol.*, **33,** 1246.
Wilgus, H., Ranweiler, J. S., Wilson, G. S. and Stellwagen, E. (1978) *J. Biol. Chem.*, **253,** 3265–3272.
Woods, K. R. and Wang, K.-T. (1967) *Biochim. Biophys. Acta*, **133,** 369–370.
Woodward, R. B., Olofson, R. A. and Mayer, H. (1961) *J. Amer. Chem. Soc.*, **83,** 1010–1012.
Wrigley, C. (1968) *Science Tools*, **15,** 17–25.

Wünsch, E. (1974) *Methoden der Organischen Chemie* (general Ed. Müller, E.), Vol. 15, Houben-Weyl, 4th edn., in two volumes. Georg Thieme, Stuttgart.
Zahn, H. and Pätzold, W. (1963) *Chem. Ber.*, **96**, 2566–2576.
Zahn, H., Brandenburg, D. and Gattner, H. G. (1972) *Diabetes*, **21**, 468–475.
Zioudrou, C., Wilchek, M. and Patchornik, A. (1965) *Biochemistry*, **4**, 1811–1822.

Index

Page numbers in italic type indicate passages with full experimental detail.